# Improving
# Your Health
# with
# Zinc

MOVIES TO GO
808 WEST SUNSET
SPRINGDALE, AR 72764
756-8660

## by Ruth Adams
## and
## Frank Murray

bunchesabooks@juno.com
(501) 872-8400
SPRINGDALE AR 72764
16618 E HWY 412
BUNCHES A' BOOKS

Preventive Health
Library
Series

**Larchmont Books
New York**

*NOTICE: This book is meant as an informational guide for the prevention of disease. For conditions of ill-health, we recommend that you see a physician, psychiatrist or other professional licensed to treat disease. These days, many medical practitioners are discovering that a strong nutritional program supports and fortifies whatever therapy they may use, as well as effectively preventing a recurrence of the illness.*

*First printing: June, 1978*
*Second printing: March, 1981*

**IMPROVING YOUR HEALTH WITH ZINC**

Copyright © Larchmont Books, 1978

*ISBN 0-915962-26-8*

*All rights reserved. No part of this book may be reproduced without permission in writing from the publishers.*

*Printed in the United States of America*

**LARCHMONT BOOKS**
6 East 43rd Street
New York, N.Y. 10017
212-949-0800

# Contents

# CHAPTER 1

# The Importance of Zinc

PROPOSED CHANGES in the national school lunch program may result in meals that do not provide adequate nutrition, according to a statement made by Dr. Peyton Davis, Director of Nutrition of the National Live Stock and Meat Board, Chicago, Illinois, in November, 1977. Dr. Davis presented comments on the proposed changes at a public hearing held by the U.S. Department of Agriculture in Denver, Colorado. The USDA administers the School Lunch Program.

In appraising the nutrient content of menus based on the USDA recommendations, Dr. Davis found:

1). Not enough zinc for any of the groups, except when a full portion of meat was served.

2). Too little iron and zinc for children one to two years of age.

3). Adequate protein but only about one-half the required calories in all age groups. Dr. Davis emphasized that when children are short of calories they tend to eat more of the less nutritious foods.

4). Magnesium, calcium, thiamine, pyridoxine and vitamin A were low in some age groups and adequate in others.

Dr. Davis' findings were obtained by calculating the nutrient content of menus containing the USDA's suggested amounts of foods and portion sizes indicated in the September 9, 1977 *Federal Register.* Meals for five age groups of children—including pre-schoolers—are planned and served through the National School Lunch Program.

The USDA is recommending changes in the school lunch program to meet more accurately the nutritional needs of children of varying ages, to bring the lunch requirements into conformity with the 1974 revisions of the Recommended Dietary Allowances, reduce plate waste and to use the program as a means of teaching nutrition to children. Of course, by the time the USDA makes a decision, the Food and Nutrition Board of the National Research Council-National Academy of Sciences, will have issued new RDAs, which the Board does every four or five years. So much for bureaucrats.

**Speaking of the possible iron and zinc deficiencies for the 1-2-year-old group,** Dr. Davis said, "These two trace minerals are exceedingly important in the diets of children because of the connection of iron with anemia and zinc with hypogonadism (abnormally small reproductive organs) along with reduced growth." He noted that iron is readily available in meat, which also helps the utilization of iron from other food sources for all five of the age groups.

He added that one way to decrease plate waste and supply larger amounts of the needed nutrients is to increase slightly the portion size of the meat that is so well received by all children. Dr. Davis said that his nutrient calculations are based on normal portion sizes and that if the portion actually served is smaller or if the child doesn't eat all of the food, there is a possibility that nutrient deficiencies could develop. He noted that a survey by the General Accounting Office showed that school lunch portions were below the size required in a number of instances.

"The School Lunch Program should supply the optimum amount of nutrients because the other two meals may well be low," Dr. Davis said. "If the luncheon meal is low and the other home meals are low, the children are in nutrient jeopardy."

**Of course, children who are deprived of essential nutrients in their developing years go on to become victims of the various zinc deficiencies and other maladies in later life.**

Adolph (Rudy) Coniglio, a Closter, New Jersey pizza store owner, has the distinction of being the first victim of **idiopathic hypogeusia** (lack of a sense of taste) whose experiences have been recorded in considerable detail, according to an article by Berton Roueche in the September 12, 1977 issue of *The New Yorker*.

Following a bout with a cold, Coniglio returned to his pizza shop, where his specialty was *pizza alla napoletana*. But the tomatoes he was using began to smell funny. He tasted them and "they tasted like garbage." A doctor prescribed some pink antibiotic pills, but there was no relief. **All of the regular food that he ate smelled and tasted like garbage.**

As he began losing weight, he went from doctor to doctor, both in the United States and in Italy (where he escaped to visit his mother and try to find out what was causing his trouble). They began prescribing more antibiotics, suggesting that he see a psychiatrist, recommending surgery for alleged tumors in the nose, etc. After seeing about a dozen doctors and a psychiatrist, Coniglio was referred to Dr. Robert I. Henkin, Director of the Center for Molecular Nutrition and Sensory Disorders at the Georgetown University Medical Center in Washington, D. C.

**Aberrations of taste and smell are not at all unusual, Dr. Henkin says. They are common symptoms of the common cold, hepatitis and pregnancy.**

"I gave (Coniglio) a drop of relatively concentrated hydrochloric acid," Dr. Henkin said. "This is a compound that normal people find extremely sour. But Rudy didn't bat an eye. He couldn't even taste it."

Dr. Henkin arranged for Coniglio to enter Bethesda clinical center, where routine tests were made: skull tests, EEG, brain scan, sinus series, a neurological examination, an ophthalmological study, copper and zinc evaluations and a tongue biopsy. The examination of tongue tissue could tell Dr. Henkin if there had been any anatomical changes in the area of the taste buds.

All of the tests were normal, except the trace-metal evaluations and the tongue biopsy. Both serum and urinary levels of zinc and copper were low. On the fifth day of his hospital stay, Coniglio was started on an oral zinc sulfate—400 milligrams a day. There was some improvement during the first two weeks. Then the tongue biopsy reports came in. Dr. Henkin found that the normal structure of the patient's taste buds was almost totally absent. Looking at photographic enlargements of the electron-microscope examination, Dr. Henkin noted that the taste buds looked "frayed, worn down, moth-eaten." Since Coniglio had been on the zinc therapy for two weeks, Dr. Henkin decided to make another tongue biopsy. The results looked like a normal tongue. By then, Coniglio had returned to New Jersey.

That was in January. In July Coniglio returned to Washington and reported to Dr. Henkin that everything was normal: he was eating well and he no longer feared that he was losing his mind. "I came back home that second time from the hospital, and I never felt better in my life.... For almost a year now, I could eat and drink and smoke and work at my place. I had a normal life," Coniglio says.

But in October he noticed smoething different. Things began to smell again. He was taking his pills but the garbage taste and smell had returned, albeit not as pronounced as at first. He returned to Bethesda and told Dr. Henkin what was happening. Dr. Henkin proceeded to laugh. To make a long story short, Dr. Henkin was so anxious to keep track of Coniglio and to make several tests that, on the patient's previous visit, Dr. Henkin had sent him home with a placebo (a harmless sugar pill) rather than the zinc supplement.

**Dr. Henkin believes that there are well over one million Americans who are suffering from some form of taste or smell dysfunction.**

8

"That's a lot of people, and problems of taste are very disagreeable problems," he says. "Eating is a basic pleasure. I would say that eating and sex are the two basic pleasures in life. We're being told now that the pleasures of sex may well be enjoyed into very ripe old age, and I hope it is true. But it is certain that eating is with us *all* the years of our life. A life without pleasure in eating—a life in which nothing tastes good and much tastes awful—would be hard to endure.

"We hope that we're on the way to bringing help to those unhappy people who are forced to endure it. Oh, there's so much ahead of us. I'm an activist and an enthusiast, and the potential in our work could hardly be more exciting. We're on the threshold of a new technology. I'm thinking beyond the treatment of hypogeusia—important as that is. I'm thinking, for example, of the day when we can order the chemistry of saliva in such a way as to prevent dental caries and gum disease. I'm also thinking of what a fuller knowledge of the physiology of taste could mean to the food industry. We know now that the taste receptors can be manipulated. Sour or bitter can be made to taste sweet. Think what the refinement and application of that phenomenon would mean to the diabetic, the obese, the hypertensive, the heart sufferer. Sweetness without sugar, sweetness without sweeteners. No more worries about saccharin or the cyclamates . . ." Dr. Henkin says.

**White spots on fingernails, which are very common among children and some adults, may be caused by deficiency in zinc and pyridoxine (vitamin B6).** So say two physicians from the Brain Bio Center at Stillman, New Jersey, in a letter to the editor of the *Journal of the American Medical Association* for April 8, 1974. The zinc deficiency could come about for different reasons. Perhaps protein is being lost in the urine and zinc is bound to protein. Or perhaps the high copper level in women's blood during menstruation may unbalance the zinc content of their blood. An acute psychotic episode may cause lack of appetite hence lack of enough zinc. Or the patient may be drinking water with too much copper in it, which unbalances the amount of zinc in the blood. Birth control pills containing estrogen also raise the copper levels of blood and lower the zinc levels.

Dr. Carl C. Pfeiffer and Elizabeth H. Jenney, M.S., point out that, if bread is enriched with iron, this, too, will diminish the amount of zinc available in our diets. Substituting soybean protein for meat in hamburgers, on a nationwide scale, is likely to increase our intake of copper, and decrease our intake of zinc, since red meat is the best source. These two researchers have found that they can remove small white spots on fingernails by giving patients zinc and pyridoxine. Large white spots must grow out as the nail grows out. This takes up to six months, but they do not reappear, so long as therapy is continued. These researchers are now doing tests on the nail spots to determine their trace mineral content, in comparison with that of normal fingernail tissue.

A laboratory experiment, reported in *The Journal of Nutrition* for January, 1974 has shown that **when you get less than optimum amounts of protein in your diet, you are likely to absorb less zinc.** The more protein you get, the healthier your zinc condition is likely to be.

Two Russian scientists report in a 1971 Russian journal that they tested levels of zinc, copper, nickel, manganese and molybdenum in people with various rheumatic diseases. **They found unusual levels of zinc and molybdenum in people with acute rheumatic diseases.** In chronic inflammatory diseases, the accumulation of most of the trace minerals varied more than in the case of acute disease. No one knows what any of this portends, but it may indicate that the trace minerals play some part in preventing or (if imbalances exist) in causing some of these conditions.

As the newspaper headlines and TV news reports caution us that we are living in an age of terrible violence so far as individual attacks upon other human beings are concerned, evidence continues to mount, indicating that what we eat and do not eat may have a great deal to do with this unprovoked violence. Now a laboratory experiment reported in the

10

September 18, 1975 issue of the British scientific journal, *Nature*, indicates that lack of a trace mineral—zinc—in the diets of pregnant animals may condemn their offspring to aggressive natures of such seriousness that they are ready to resort to violence at the slightest provocation.

**Scientists know that lack of zinc can create smaller brains with fewer brain cells in offspring of the deficient mothers.** Diets of children in some parts of our country have indicated very low levels of zinc, causing symptoms which might pass unnoticed at a medical examination. Lack of appetite is one. Studying laboratory rats born to zinc-deficient mothers, a group of researchers at the University of North Dakota and Human Nutrition Laboratory of the USDA in Grand Forks, found that these deficient offspring were less able to withstand stress. When they were placed with well-nourished rats and all were subjected to stress, the deficient ones bit, fought and assumed aggressive attitudes whereas the well-fed ones did not. The scientists say they have no proof that the same things occur in human babies, but they think "a closer look is warranted."

A Georgetown University research team has isolated a protein which is contained in saliva and which seems to be responsible for the growth and differentiation of taste buds. This protein contains zinc. **When zinc is missing or not present in large enough amounts, taste is markedly impaired.** People who do not get enough zinc suffer not only from defective taste buds, but also not nearly enough of the taste buds and a disorder in those that remain.

A survey of Head-Start children revealed that half of them were deficient in zinc. Of 4,000 patients going to the university clinic complaining of taste disorders, 10 per cent had deficiency in zinc caused by undiscovered cancers. It seems that cancer cells may collect whatever zinc is available in the body. During the first three months of pregnancy when the growing fetus is developing rapidly growing tissues, zinc may be lacking for the mother, who then may develop peculiar eating habits and complain of distorted taste.

11

Deficiency in zinc can also be caused by underactive thyroid gland, liver disease, lack of proper absorption in the colon or a large intake of foods that contain phytate. We would add that wholegrains contain phytate, but when it is made into leavened bread (with yeast) its powers for preventing the absorption of minerals are largely overcome.

The Georgetown professor, Dr. Robert I. Henkin, whom we mention a number of times in this book, said that **at least 20 per cent of people who complain of taste disorders cannot absorb zinc when it is given to them, even though they have no trouble absorbing all other minerals.** Doctors don't know why. One obvious reason, of course, is the overprocessing of our grains and sugars.

Says Dr. Henkin, "In the farthest reaches of our imagination, I don't think we have any idea how important and how widespread zinc deficiency problems are. When you consider the small amount of zinc in the diet, and that it is mainly present in costly, high-protein foods, I think we will begin to see much more zinc deficiency now that we can identify it."

*Nature* reported in its June 18, 1971 issue that laboratory animals given a supplement containing 22 parts per million of zinc were able to withstand a cancer-causing drug which caused tumors in a second group of animals which did not get the zinc supplement. The article concludes that giving zinc as a dietary supplement "seems to exert an inhibitory effect on tumor formation."

A relationship between zinc in adrenal glands and the amount of cholesterol in those glands was discovered by a Scots researcher and reported in *Proceedings of the Nutrition Society*, September, 1972. Laboratory rats kept on diets deficient in zinc had more cholesterol in their adrenal glands than those kept on diets containing plenty of the mineral.

Absence of the sense of taste was noted in some children who were found to have low zinc levels, during a survey of the amount of zinc in their hair. Measuring trace minerals in hair is an acceptable way of determining the body store. Ten children out of 338 apparently normal children were found to have low amounts of zinc in their hair.

**Five of these had almost no sense of taste.** (Doctors call this hypogeusia). They also had poor appetites. They were given zinc supplements and, after three months, appetites returned to normal and sense of taste returned as well. Four scientists report these experiments in *Pediatric Research*, Vol. 6: page 868, 1972.

In 1968, Dr. William B. Bean of the Department of Internal Medicine at the University of Iowa said that there is some evidence that **diets deficient in zinc may set the stage for rheumatoid arthritis.** Chicks fed diets deficient in zinc develop bone enlargements and deformities that resemble human arthritis. And deficiency in zinc causes an increase in congenital deformities in animals.

A Greek scientist (Research Laboratory of Physiopathology of Animal Reproduction, Athens) reported in a letter to the editor of *The Lancet*, December 9, 1972 that he had noticed in himself an unexplained result of taking zinc supplements.

Says Dr. F. N. Demertzis, "Working recently with zinc, I was impressed with the effect of zinc supplements on apparently healthy animals. So I decided to take zinc myself. The result was surprising. Firstly the long hairs of the eyebrows (a sign of the aged) disappeared, and new, short and thin adolescent-like eyebrows took their place. The hair became more healthy and shining, its color darker, and every trace of dandruff disappeared. In the comb in the morning there was not a single hair anymore. Finally the greasy skin (full of acne at the time of my adolescence) became dry and better than I had ever had it. After my experience, the same effect was noted in three other people I know who took zinc."

According to *Today's Health* for September, 1974, a du Pont chemist theorizes that **zinc may help vitamin C in preventing colds.** He experimented with zinc ascorbate—a compound of the trace mineral zinc and vitamin C. He found that this compound destroyed 12 rhinoviruses—the bugs

13

that cause colds. Was it the zinc or the vitamin that did the trick? He did not know, so he analyzed several commercial preparations of vitamin C and found that they contained a bit of zinc, as well. If the brand of vitamin C you use contains some zinc along with its vitamin C, this would appear to be an added benefit.

Two researchers at the Oklahoma City Veterans Administration Hospital injected laboratory mice with enough alcohol to kill them, or almost kill them. Some of the mice had been given large amounts of vitamin C before the injection. All of these animals survived. The rest died.

Another group of mice, injected with alcohol, was given zinc before the injection. Of these animals, 90 per cent survived. Of those mice which got no nutritional support less than one-third survived.

In another experiment, rats were injected with alcohol at frequent intervals for four weeks. They became what, in human beings, is described as chronic alcoholics. An hour after each injection the alcohol was measured in their blood. **In animals which had been given vitamin C before the injections, the level of alcohol was considerably lower than in those which got no vitamin. The same was true of zinc.**

The researchers say, as researchers always do, that their experiments only partially support the theory that the same might be true of human beings given enough zinc and vitamin C. Research is continuing, they say.

We say, research or no research, alcohol or no alcohol, all of us should be getting far more vitamin C than we usually get and far more zinc than we get in the average American diet. Alcohol is a poison. Enough vitamin C is a protective shield against many poisons. Now it appears that alcohol is one of these. Zinc is used by the body in dealing with all carbohydrates. This is why it is so essential for those with low blood sugar or diabetes. There is no zinc in any alcoholic drink, just as there is no zinc in sugar or white bread. So increasing your intake of these substances is bound to leave

14

you deficient in zinc.

Statistics on the sales of hair dyes indicate rather conclusively that many millions of Americans are worried about the color or lack of color in their hair. Most troublesome of these worries is white hair or "gray" hair, which is never really gray but gives that appearance due to a mixture of white and dark hair.

Well, doesn't white hair make you look old? Unfortunately, our word associations with symptoms of aging make such a connotation inevitable. "She was an old, white-haired lady," we say, or "He's aged greatly. See how white his hair is." There seems to be some justification for the assumption that anyone with white hair is older than someone whose hair retains its natural color. On the other hand, there are many many people these days whose hair becomes entirely white or almost entirely white at a relatively early age, yet they survive in apparent good health and suffer few other symptoms of premature aging. Why?

There is a nagging suspicion that diet must have something to do with it. But what items of diet? What deficiencies? In animal experiments where living creatures can be confined and restricted to very closely measured diets, hair color can be changed by diet. But it is next to impossible to submit even small groups of human beings to such rigorously controlled experiments. So how can we know how much bearing such experiments have on human hair and human experiments? We can't know, but we can make some educated guesses.

An article in *Environmental Research*, Vol. 6, 247-252, 1973, tells of a study of trace minerals in human hair from residents of Port Arthur, Texas, an industrial city where one might expect considerable accumulations of trace mineral pollutants. These may show up in hair samples. No one knows quite why this should be so, but some scientists believe that this may be one reason why we have hair—so that it can act as an excretory organ for toxic trace metals. Lead, for example, collects in hair. Children with diagnosed lead poisoning have relatively large amounts of lead in their hair. Perhaps this is one of the body's ways of getting rid of this poison, as the hair eventually falls out.

Drs. Edwin A. Eads and Charles E. Lambdin of Lamar University tell us that **a comparison of the zinc and copper content of hair showed a strong connection with the color of the hair**. The zinc to copper ratio is highest for dark hair and decreases with the increase in lightness in hair color. The "gray" hair of both men and women shows lower zinc-copper ratio than that of the non-gray hair of younger people.

One subject, aged 60, had gray hair with a zinc-to-copper ratio of 5.4. Another, at the age of 44, had a ratio of 8.6, while black hair, brown hair and red-brown hair went as high as 15.8 in these two trace minerals. **A significantly higher level of zinc was present in every person with darker hair.** "It is reasoned," say the authors, "that zinc may be implicated in the production of the melanin pigments in the hair."

In the classic book on trace minerals, *Trace Elements in Human and Animal Nutrition*, Dr. E. J. Underwood tells us that, in sheep, the color of their hair is intimately related to the copper in their diets. In sheep whose black wool is an inherited characteristic, lack of copper turns the hair white long before there is any other symptom of anemia or other copper deficiency ailment. Sheepmen know they can produce wool which is alternately black, then white, then black again, by varying the amounts of copper in the diets of their animals.

**The health of everything associated with skin appears to be related to the zinc content of the diet.** Fingernails of 18 normal healthy people were found to contain from 93 to 292 parts per million of zinc. They averaged 151 parts per million. The hair of 46 people who were tested ranged from 92 to 255 parts per million, with an average of 173.

**In some parts of the world diets are very low in zinc and people eating such diets are frequently dwarfed.**

16

The hair of these people has only 54.1 parts per million of zinc. Giving them ample amounts of zinc produces hair with normal amounts of zinc in it. Dr. Underwood tells us that there is some correlation with age in the zinc content of hair, but he could not uncover any evidence that hair color was related to zinc.

Dr. Carl Pfeiffer, previously mentioned, believes that Americans in general get too much copper. In *Medical World News* for April 14, 1972 he says that people who live in suburbs with their own well drink acid water, generally, which erodes the copper from their water pipes. The Public Health Service has set one part per million as the desirable amount of copper for drinking water. Dr. Pfeiffer says that his patients bring in samples of drinking water with five times that much. **Schizophrenics have too much copper in their blood.** Dr. Pfeiffer, who treats schizophrenics with megavitamin therapy and other dietary measures, gives his mental patients zinc which balances the too-great amount of copper.

He is very definite in his recommendations for never drinking the first water that comes out of the pipes in the morning, if you have copper pipes. This water has been standing in the pipes all night and has probably picked up more copper than you want for your breakfast coffee substitute or tea. Draw your breakfast water the night before, says Dr. Pfeiffer, so that it will have less copper.

A group of California scientists has uncovered a curious element present in human breast milk which is not present in cow's milk. This suggests that the finding may have special meaning for children brought up on formula made from cow's milk. Are they lacking something, right from birth on?

They may be, and that something may be zinc. Four biologists from the University of California in Davis reported in *The Lancet* for January 22, 1977 that they have found in human breast milk a "zinc binding ligand"—that is, a substance which helps the baby drinking the milk to absorb zinc.

They believe, they say, that the presence of this substance indicates that **human breast milk may be more important to the health of infants than hitherto suspected.** For it may be that this substance allows the human infant to absorb this essential trace mineral during the period immediately after birth when any normal mechanisms for absorbing zinc may be absent or immature.

They have found the same substance in milk from rats and they have found a similar substance in the intestines of mature rats, but not in newborn rats. **All this seems to indicate that zinc is indeed extremely important for good health and that babies brought up mostly on cow's milk may be missing out on zinc entirely, if they do not have the ability to absorb whatever zinc they get from other food.**

We know that zinc helps infants to grow and that diets in which there is almost no zinc tend to produce dwarfed individuals who do not mature normally.

Says an article in *Medical World News* for December 27, 1976, "If any infant or young child seems to have a poor appetite and shows signs of lagging growth, in the absence of any other explanation he may be suffering from zinc deficiency." They go on to review a recent experiment at the University of Colorado in which they gave a group of 34 babies two different formulas, one of them with considerably more zinc than the other.

They found that after six months the babies who had been getting the larger amounts of zinc weighed more, especially in the case of boy babies. Testing the zinc in their blood they found that those infants with more of the trace mineral in their blood were those who had shown the most growth. Once again, this happened only with the boys. The girl babies showed about the same level of zinc in the blood of those who put on more weight and those who put on less.

All babies had been drinking formulas whose zinc content was well known. After six months all were given cow's milk and the boys who had done the most growing "continued to hold their gains." One result of this experiment is that some of the baby formula manufacturers have begun to add zinc to their formulas.

"Plenty of babies aren't on formulas," said one of the Denver researchers. "If they're on fresh cow's milk and getting enough of it they're probably doing all right. But there's a growing body of evidence—much of it from the Denver studies—that there's a zinc deficiency problem in preschool children.... **There may be as many children with too little zinc in their diets as with iron deficiency.**"

The reason why commercial formulas lack zinc is because the cow's milk which is used in them is diluted so that its protein content will more nearly approximate that of human milk. This dilutes the zinc content as well. And, we would add, the content of all the other minerals and trace minerals as well.

**The ability of zinc to fight infections** came in for some helpful comments at a recent conference on zinc in health and disease. A large part of the time was taken up with discussion of just how zinc induces the body to fight off dangerous invaders.

**Prostatic fluid (the fluid in the male prostate gland) contains more zinc than any other human tissue or secretion, apparently indicating the great importance of this trace mineral to the health of this organ.** Several biologists from Washington University have discovered, they say, an antibacterial factor which they named PAF for Prostatic Antibacterial Factor. This substance does indeed kill germs and the Washington University scientists found that its power against bacteria is in direct relation to the amount of zinc the fluid contains.

Examining 61 specimens of prostatic secretion from 15 men with chronic prostate bacterial infections, they found an average of only 50 micrograms of zinc per milliliter, whereas 65 specimens from men free from infection averaged as much as 448 micrograms which is about nine times more zinc!

Other essential substances in the prostatic fluid were deficient, too, in the men with bacterial infection.

A scientific adviser to the U. S. Army Medical Research Institute of Infectious Diseases revealed that levels of zinc in the blood drop sharply when an infectious process starts—or even before it starts. The body apparently has some kind of warning system so that it knows when an infection is beginning. And the zinc is apparently put to work at once to help in the fight against it.

The activity of zinc is apparently closely related to the work of the blood corpuscles which fight off invading bacteria by engulfing and destroying them. A Wayne State University team told of their observations that some patients with sickle cell anemia are deficient in zinc and are also known to be especially susceptible to infection. Zinc deficiency in male mice results in disordered lymph glands. These glands help in the body's defense against bacterial invaders.

Biologists from the University of Arizona told of experiments in which they tested the ability of zinc to defend animals against damage from silicon. The inflammation which results from exposure to silicon particles results in a disorder called *fibroplasia in silicosis* (an overgrowth of fibrous tissue). The scientists dusted a group of rats with silica particles and gave half of them zinc, thinking that the trace mineral might reduce the amount of damage done. It did. Says *Medical World News*, "The zinc treated animals' lesions were reduced about 50 per cent compared with the untreated animals!..."

The Army physician referred to above confessed that he was reluctant to advocate automatically "putting everyone on high levels of zinc." He said that the body apparently is very clever about distributing its zinc stores in the liver or the blood or wherever it is needed and "the location of the zinc may be more important than how much there is."

However, it seems to us that there is ample evidence that the vast majority of present-day Americans must be deficient in zinc, since they have been eating all their lives diets from which zinc is almost completely removed. In any event, it seems that the more research being done on zinc the more we discover that its effects are wholly beneficial.

Nutrition experts used to believe that it was impossible for human beings to develop deficiency in zinc, since it is so widespread in food and water. But now we know differently, says an editorial in *The Lancet* for February 10, 1973. **Zinc deficiency has been found in badly nourished nations:** zinc disappears rapidly from the body under certain kinds of stress, and zinc can also be used for various healing purposes.

**Zinc is an essential mineral, needed for bone growth.** Lack of zinc in national diets has produced a dwarfed condition along with failure of sex organs to develop. When these individuals were given zinc supplements they began to grow and sex organs matured. The deficiency in zinc was caused not so much by lack of the mineral in food as by the fact that the diet consisted almost entirely of cereal in which phytic acid was abundant. This element causes the minerals in food to be unabsorbed.

Studying amounts of zinc in hospital food, several researchers found that "good quality" hospital diets gave an average of about 7 to 16 milligrams of zinc daily. Foods rich in protein were much better sources of the mineral than refined carbohydrate foods. Because such a large part of modern diet consists of refined carbohydrate foods, there is a great possibility that many modern diets are deficient in zinc.

**Zinc seems to help in healing wounds,** some kinds of wounds at any rate. Some researchers believe this may be because the hospitalized patients were short on zinc to begin with, so all the zinc supplements are doing is to restore the normal values. How much zinc to give to correct a deficiency or to spur wound healing is not known. There seems to be no indication that it is harmful.

One reason why giving zinc may be helpful in getting wounds to heal is that wounds, bone fractures and operations cause the body to lose large amounts of zinc in urine. Dr. Gordon S. Fell of the Royal Infirmary, Glasgow, said that any disease or condition of starvation which causes a loss of muscle (untreated diabetes, for example) also causes zinc to be excreted. Skeletal muscle contains about 63 per cent of all body zinc. The total losses are large, Dr. Fell told the Ninth International Congress of Nutrition in October, 1972, and could lead to zinc deficiency in severe cases. Other nutrients as well are excreted under such conditions: nickel, potassium, nitrogen and magnesium.

"There is ample evidence," says Dr. E. J. Underwood in his massive volume, *Trace Elements in Human and Animal Nutrition*, "**that the zinc content of plants can be influenced by the soil type and the fertilizer treatment**. In a limited Australian study, the zinc content of wheaten grain was markedly influenced by fertilizer treatment."

Consider that statement in the light of the incessant propaganda from the Food and Drug Administration that fertilizers and soil condition have absolutely no influence on the mineral or trace mineral content of the food grown on it! Dr. Underwood is a world-renowned authority on minerals, his book is published by Academic Press in New York and London.

Dr. Underwood tells us that the body organ which concentrates most zinc is the muscular coating of the eye. Scientists do not as yet know why zinc is apparently so important to the eye or what function it plays in sight. Then there is the male prostate gland of which we have already written.

For some reason, not understood, the levels of zinc in blood vary from one geographical region of the United States to another. Could this be because soil or drinking water in some parts of the country lack zinc? No final explanation has been made, but, in view of the fact that the amount of zinc available for human beings is "marginal," perhaps we should become concerned about this discrepancy.

Dr. Underwood tells us that white flour milled from the kind of wheat used to make bread contains 7.8 parts per million of zinc compared to 35 parts per million in the original wheat. Some researchers have recommended strongly that bread and processed cereals be "enriched" with zinc as they are now enriched with iron.

Says Dr. Underwood, ". . . if a rapid rate of wound healing is taken as a criterion of adequacy, many dietaries must be considered inadequate or at best marginal in zinc, even though other signs of zinc deficiency, as seen in animals, are not apparent."

It's hard to imagine a more difficult or discouraging disease to have or to treat than one called *acrodermatitis enteropathica*, which we discuss elsewhere in this book. Fortunately it is rare. Unfortunately it is apparently sometimes inherited. It brings terrible blistering sores on the skin, sores on the mucus membranes of mouth and throat, diarrhea and loss of hair. Usually it occurs in infants when they are taken off breast milk and given formulas.

Drugs which have been used are sometimes successful. If they are not, the baby must be kept on breast milk, no matter how this is arranged. If untreated, says an article in *New England Journal of Medicine* for April 24, 1975, the disease is usually fatal, though some victims have survived and have apparently outgrown this inherited defect, although we have been told that inherited defects usually persist throughout life.

Quite recently zinc has been used to treat this disorder and patients have responded "dramatically." The *Journal* article describes a patient, this one a 22-year-old woman who developed the disease when she was only three months old. When she was 1½ years old, she was at the point of death when her life was saved by drugs which she had to take every day. For some reason the drug increased the amount of zinc in her blood. Several years ago this observation led to doctors giving the trace mineral zinc to see what the effects would be.

In this case, as in earlier ones, the effects were startling and gratifying. This woman had had two pregnancies which ended in spontaneous abortions. Then she had a stillborn defective child. Since then she has taken The Pill.

The doctors at a Colorado hospital gave her first the drug which usually controlled her condition. Then they withheld it until ominous symptoms began to appear. Blisters began to form on her knees, her mouth tissues became sore and she began to feel ill and depressed.

They gave her zinc sulfate—220 milligrams three times a day—and she responded rapidly and dramatically. Within 24 hours she felt well. Within four days her skin blisters and other symptoms had completely disappeared. Levels of zinc in her blood became normal. The doctors stopped the treatment after 10 days and she remained well without any treatment. But within five weeks symptoms had returned and the zinc levels in her blood were once again low.

**On zinc therapy—100 milligrams this time, given in two daily doses, she improved once again—showing improvement within 24 hours.** The doctors believe, they say, that her response to this trace mineral indicates that this disease is very much involved with the metabolism of zinc. Either the patient cannot absorb enough zinc to keep her healthy or she loses it before it can do her any good. The amount finally needed to keep her well is, they comment, only a little above the recommended daily allowance for zinc, which is 15 milligrams. The 100 milligrams of zinc sulfate contain about 22 milligrams of zinc.

Health seekers sometimes ask what vitamin or mineral they should take to get rid of some one symptom which they describe. As you can see from the above story, there is no such thing as one symptom which occurs because of deficiency in some essential food element. Zinc, like other minerals and vitamins, too, is used in so many locations in the body that a deficiency is bound to show up in many places— skin, mouth, mental health, reproductive organs, bones.

In his book, *The Best Health Ideas I Know*, Robert Rodale says **there is lots of evidence that we need far more zinc than the present estimate for a daily allowance.** One of the main reasons may be that the zinc content in our food is declining all the time because of the use of modern commercial fertilizers which provide no zinc. Yields on grain are increasing because of all the fertilizer our farmers use, but the zinc content of soils is dropping, not only in this country but also in many other parts of the world.

He tells us that a group of Colorado children were recently found to be quite short on zinc, as indicated by slow growth rates and severe deficiencies in the senses of taste and smell. Another specialist in the field points out that vegetarians may suffer most from zinc deficiency. This mineral is needed for the body to use protein. It is most abundant in high protein foods like meat and fish. So vegetarians who use no animal products may lack zinc, even though they get enough protein in their vegetarian foods. They cannot use all this protein without enough zinc to process it. And vegetarian foods may be especially lacking in zinc due to lack in the soils in which food is grown.

Two Midwestern scientists, Dr. Donald F. Caldwell and Dr. Donald Oberleas, believe that **as many as 80 per cent of all Americans may be deficient in zinc**. Our bodies lose zinc in excretions when we have colds, also in perspiration. Zinc is lost, too, when foods are processed. Even canning and cooking cause some loss of the mineral.

Drs. William Strain and Walter Pories found that **people suffering from hardening of the arteries were deficient in zinc**. Giving zinc supplements brought some dramatic benefits to such patients. Two Tennessee doctors reported healing of gums after tooth extractions when zinc was given.

Dr. Jean Mayer reports that **zinc helps to recover from fatigue caused by exercise and helps undo the effects of alcohol**. More than 80 different enzyme systems in the body depend on zinc as one of their ingredients—about three times more than are dependent on magnesium, another trace mineral. Enzyme systems are those groups of nutrients— proteins, hormones, vitamins and minerals—which bring about chemical changes in our bodies, allowing us to utilize food and build it into our cells or use it as energy.

Zinc therapy is not an overnight therapy. As is the case

25

with any nutritional therapy, it works slowly. It may take quite a long time to undo the damage done over many years by inadequate diet. Zinc is a common mineral, which cannot be patented or sold as an expensive drug. So drug companies are not much interested in promoting it. They can't make much money out of it. Doctors get much, if not most, of their information on therapy from drug salesmen and drug advertising.

And most doctors, like most official spokesmen on nutrition, believe that no patient is short on zinc unless he comes in half-dead from well-recognized symptoms of zinc deficiency. Not many of us in this country ever show such definite symptoms. But this does not mean we are getting optimum amounts of zinc for good health.

One specialist in zinc metabolism believes adults should be getting up to 25 milligrams daily. Dr. Harold Rosenberg in his book, *The Doctor's Book on Vitamin Therapy*, recommends up to 30 milligrams. Bone meal contains zinc in natural form, but not a great deal in terms of the potencies we are talking about above.

To reiterate, **here are some possible causes of zinc deficiency** as they are listed in *Trace Elements in Human Health and Disease, Volume I, Zinc and Copper*, edited by Ananda S. Prasad, and published by Academic Press.

Too much phytate in the diet as in cases of people living on unleavened bread with not much else in the diet. *Pica*—the eating of peculiar things like clay and laundry starch. Taking the drug EDTA to eliminate some toxic trace mineral like mercury.

Alcohol increases urinary excretion of zinc. Malfunctioning of pancreas. Operations which remove part or all of the stomach. Diverticulitis. Diseases of the lining of the intestine. Cirrhosis of the liver and other liver diseases. Kidney failure. Lack of ability to absorb minerals from the intestine. Nephrotic syndrome. Renal tubular disease. Dialysis. Hemolytic anemias. Cancer of various kinds. Psoriasis. Burns. Parasitic infections (intestinal worms, for example).

**Reasons for zinc deficiency caused by medical treatment are these.** Antimetabolic drugs. Antianabolic drugs. Chelating drugs. Intravenous feeding. Surgery.

**Here are the reasons for zinc deficiency caused by inherited defects or congenital defects.** Hemolytic anemias. Pancreatic defect. Intestinal defect. Renal tubular defect. Diabetes mellitus. Congenital absence of thymus gland. Mongolism. Acrodermatitis enteropathica.

Infectious diseases, collagen diseases (arthritis), tissue necrosis. Pregnancy.

We would add to this list: Schizophrenia, epilepsy, acne, heart disease, depression, eye health, prostate gland disorders, lack of sense of smell and taste . . . and many more.

One hundred years ago the average man ate a great deal of bread—tough, brown, chewy bread made of wholegrain flour. Many people, especially poor people, lived almost entirely on bread. They got relatively enormous amounts of zinc from the bread, as well as from the seeds and nuts which they had at meals.

Today children are brought up on white bread, hot dog and hamburger rolls, English muffins, along with enormous quantities of cakes, cookies and pastries all made from white flour. They eat for breakfast cereals from which all the trace minerals have been removed. When they grow up the pattern remains the same.

At a hearing of the Senate Select Committee on Nutrition and Human Needs, Dr. Jean Mayer said, speaking of the sugar-coated cereals that are so popular among children, "According to testimony of many young mothers, these are often eaten like candy, without milk. In spite of their being enriched with some vitamins and iron, the total effect is one of inadequate nutrition (deficiency in particular in trace minerals) —there are suggestions that zinc deficiency may be appearing among U. S. children, including middle and upper socio-economic class children; chromium deficiency may be a factor among the elderly."

27

In 1973, testifying before a Senate Select Committee, Dr. Walter Mertz, Chairman of Human Nutrition Institute of the U. S. Department of Agriculture, said, "We have not yet learned to understand the optimum requirement for all essential trace nutrients. Therefore, if we fabricate our own foods, we must accept that our knowledge is incomplete and therefore it is entirely possible that our fabricated foods are inferior in quality to that of the more wholesome products... we certainly have an example in zinc nutrition. **In the past 5 to 10 years evidence has accumulated that the zinc nutrition status of a proportion of our older population is not optimal as shown by very good effects of increasing their zinc intake."**

He went on to describe **one of the commonest symptoms of zinc deficiency—lack of or impairment of a sense of taste**. One survey showed that 8 to 10 per cent of supposedly normal children from middle and high income families were markedly deficient in zinc, as shown by the fact that their taste sensation was deficient, and they lacked appetite. Lack of appetite is very common among older folks. And often it is accompanied by the complaint, "I just don't seem to be able to taste anything any more."

Dr. Mertz added that, **"The daily intake of zinc may furnish or be slightly below the human requirement**. Recent research already has identified population groups that are at risk of marginal or deficient intake. Although the available data are still limited, they alert us to the possibility that marginal zinc deficiency may not be uncommon in the United States. This recognition led to the fortification by the Food and Nutrition Board subcommittee. It is already implemented by most of the infant food manufacturers."

A number of years ago millers were required to restore the iron which they remove in refining flour—but no one ever made them restore all the other trace minerals they remove—like zinc. After 50 years or so of this kind of eating, it seems perfectly possible, in fact, probable, that prostate gland disorders and the other maladies reported in this book may be caused by this kind of diet. How many years will it take our learned researchers to discover how much this absence of zinc has contributed to the rising incidence of diabetes and related disorders, as well as prostate troubles?

"Refining of raw cane sugar into white sugar removes most (93 per cent) of the ash, and with it go the trace elements necessary for metabolism of sugar: 93 per cent of the chromium, 89 per cent of the manganese, 98 per cent of the cobalt, 83 per cent of the copper, 98 per cent of the zinc and 98 per cent of the magnesium. These essential elements are in the residue molasses, which is fed to cattle," says Dr. Henry Schroeder in *Trace Elements and Man.*

He also said that, "For ourselves, we avoid the refined flours and all sugar-containing foods, using wholewheat, brown rice (which also loses elements in being polished white), dark brown sugar with a high ash content, and all the natural sugar: corn, maple, honey, fruit juices, etc. Much 'raw' sugar imported into this country is considerably refined, because of excise taxes, and sugar with a high ash content is difficult to find. Brown sugar sticky from molasses is suitable; any sugar which flows freely is probably partly refined, regardless of color."

Nature does not make mistakes. Foods like cereal grains which have sustained human beings through thousands of years, have been successful as the mainstay of the diet chiefly because they contain everything that is needed to nourish us. Take away one trace element like zinc and you have destroyed the natural nourishment which this food provides. Take away all the trace minerals along with the B vitamins and vitamin E (as millers do in refining flour) and you have laid the groundwork for epidemics of deficiency disease which is what we are now witnessing in the Western world where such foods are still the basis of our meals.

Other foods than wholegrain cereals which contain appreciable amounts of zinc are these: meat and fish, especially shellfish, all seed foods (in their natural state—not refined), nuts, wheat germ, bran and brewers yeast.

29

There's one other reason to worry about getting enough zinc—the amount of the toxic metal cadmium that modern technology has brought into the world. Many industrial operations release cadmium into air and water and into products which we use in our homes. The more cadmium you get, the more zinc you need to counteract the bad effects of the cadmium.

Zinc is also available in supplements at your health food store, as are all the fine wholegrain cereals and cereal products, the unprocessed, unrefined flour, the good bread, the seeds and nuts which are rich storehouses of zinc.

Although zinc has long been officially recognized as an essential nutrient for man, it was not until the 1974 Recommended Daily Dietary Allowances were issued that we knew how much of this trace mineral we should be getting. The recommendations from the National Academy of Sciences are:

| | |
|---|---|
| Infants (to five months) | 3 mg. |
| Infants (five months to 1 year) | 5 mg. |
| Children (1 to 10) | 10 mg. |
| Males and Females (all ages) | 15 mg. |
| Pregnant Females | 20 mg. |
| Lactating Females | 25 mg. |

mg. = milligrams

# CHAPTER 2

# Your Need
# for Zinc

WHEN WE THINK OF ZINC—which is a bluish-white substance resembling magnesium—we are apt to think in terms of galvanized sheets, battery cells, roof coverings and a variety of industrial applications. But the trace mineral zinc is an important cog in the complex human nutritional machinery. And many competent nutritionists, such as Dr. Jean Mayer of Harvard University, believe that many people in the United States have a zinc deficiency.

In testimony before the U.S. Senate Select Committee on Nutrition and Human Needs, April 30, 1973, in Washington, D. C., the following exchange was made between Senator Richard Schweiker (R., Pa.), a member of the committee, and Dr. Walter Mertz, Chairman, Human Nutrition Institute, U.S. Department of Agriculture:

DR. MERTZ. It is my experience, and I have no logical explanation for it, that whenever a population becomes more well-to-do that there is a trend toward more fancy foods, there is a trend for eating increasing proportion of the meals outside the house. Now, a poor population is more or less forced to live with very little processing. We have not yet learned to understand the optimum requirement for all essential trace nutrients. Therefore, if we fabricate our own foods, we must accept that our knowledge is incomplete and, therefore, it is entirely possible that our fabricated foods are inferior in quality to that of the more wholesome products.

SENATOR SCHWEIKER. In your statement, Doctor, you refer to the National Nutritional Survey which finds there is widespread iron deficiency anemia in the country. I know in an affluent society a finding like this comes as a great shock to many people. Since we have had iron fortification policy in this country for many years, why, in your judgment, was the fortification policy inadequate, and what shall we do to correct it?

DR. MERTZ. The fortification policy was less than a full success because when we instituted it, we did not have enough basic knowledge about the availability of different iron compounds. At that time we thought that any iron salt is equal to any other iron salt. In the meantime we have learned that there are certain iron compounds which are very poorly available and others much better available. We have not incorporated this knowledge into our enrichment program of years ago. Here, again, in the past 5 to 10 years, basic nutrition has produced knowledge of certain iron compounds that are available to man and that can be incorporated into foods and that will hopefully improve the enrichment program.

SENATOR SCHWEIKER. You refer to the fact, in your statement, that people need less calories and eat less in an industrialized society. Also, that in such a situation they are more likely to end up deficient in various micronutrients. Can you give us another example besides iron?

DR. MERTZ. Yes. I would say that we certainly have an example in zinc nutrition. **In the past 5 to 10 years there has accumulated evidence that the zinc nutrition status of a proportion of our older population is not optimal as shown by very good effects of increasing zinc intake.** For example, in hospitalized patients. We are now seeing new evidence that ties the zinc nutritional status to the impairment of taste acuity, which is an extremely important

32

factor, particularly in children. Last year evidence was produced indicating that **approximately 8 to 10 per cent of a number of supposedly normal children from middle—or high-income neighborhoods examined were markedly zinc deficient, as evidenced by poor taste acuity, poor appetite and so forth**. (End of quote).

About 100 years ago a researcher named J. Raulin showed that zinc is essential for the life of a small organism. Not until 1926 did we know that higher forms of life need it also. **Eventually we learned that zinc is necessary for a wide range of processes in living cells**. An average 150-lb. man has in his body only about 1 or 2 grams of zinc, which is about half the amount of iron his body contains; 15 times more than his copper supply; 100 times more than the manganese his body contains.

Vitamin A is a fat-soluble vitamin that is stored in the liver. It is essential for seeing in the dark and in the half-dark. A number of years ago, scientists discovered that some people with liver disease could not see after dark. When they were given vitamin A, even in large doses, they still could not see, indicating that the vitamin A stores in the liver were not getting in the blood. So the scientists decided that something else must be needed to release the vitamin A from the liver.

Over the intervening years, experimenting with animals, scientists have discovered that the essential nutrient is zinc. Giving zinc along with vitamin A has produced good results. In some cases, just giving zinc has resulted in releasing vitamin A from the liver and making it available to the body. Many of us are deficient in vitamin A as well as in zinc. According to an article in *Science,* September 1973, zinc may even be able to perform some of the duties of vitamin A as far as the eyes are concerned. So, perhaps, the two nutrients are somehow interchangeable. In any case, it is wise to get enough of them in your daily diet or in food supplements. For supplements, your health food store has many formulas.

The following table shows the concentrations of zinc in human organs. The figures are for milligrams per kilogram:

| | |
|---|---|
| Liver | 141 to 245 |
| Kidneys | 184 to 230 |
| Lungs | 67 to 86 |
| Muscle | 197 to 226 |
| Pancreas | 115 to 135 |
| Heart | 100 |
| Bone | 218 |
| Prostate | 520 |
| Eye, retina | 571 |
| Eye, choroid | 562 |

It is possible to raise the zinc levels in various organs and parts of the body by taking zinc supplements. If one wants to theorize on which parts of the body seem most dependent on a goodly supply of zinc, it is wise to take a look at where the trace mineral is concentrated. For example, the male prostate gland contains more zinc than any other organ of the human body—102 micrograms per gram—almost twice as much as the liver and kidney. All body muscles—including the heart—store zinc, indicating that this mineral is apparently very important to the healthful operation of muscles and heart. The pancreas, lung, spleen, brain, testes and adrenal glands also contain appreciable amounts of zinc.

# CHAPTER 3

# Are You Short on Zinc?

A CRYING BABY, its head completely bald, its face, eyes, nose, mouth and cheeks disfigured with hideous red blotches and blisters. A woman whose occupation is flower arrangement who cannot smell. A man whose job is tea-tasting, who has lost his sense of taste. What have all these sufferers in common? Would you believe it is just that they all lack the trace mineral zinc? Would you believe that by giving them zinc in fairly large daily doses, a doctor can, in many cases, bring a final end to these diverse conditions?

**Many otherwise normal people have lost their sense of taste.** Others have lost their sense of smell. There are others whose senses of taste and smell are partially disabled, so that they have almost no ability to taste or smell, and still others whose tastebuds and sense of smell are so disordered that everything tastes and smells unpleasant to them.

By itself, such a disability seems not too serious, except that it might be life-threatening if you were unable to smell an approaching fire. And it could be most upsetting if you were employed as a tea taster or a connoisseur of perfumes. But other eventualities occur as a result of these disorders. The individual who can neither smell nor taste can certainly not appreciate food, hence will probably lack appetite. Since the

35

sense of smell is closely related to the sense of taste, this, too, could easily produce disinterest in food so serious as to be life-threatening.

According to *Medical World News* for October 25, 1974, there are many causes of these disorders. Some thyroid disorders may destroy so much of the sense of taste that such patients use immense amounts of salt over their food, since they cannot taste it. This could also be one reason for people eating far more sugar than they should have. If a sense of taste is blunted, the patient feels he must go on adding sugar until he can taste it.

Head injury or burns, badly fitting dentures, operations on the larynx and several kinds of viral disorders can also produce taste and smell anomalies. Diseases of the nerves and muscles, diphtheria and encephalitis, strokes and sometimes the flu can produce transient or even more permanent distress in this area. Drugs can also produce such disabilities.

Dr. Robert I. Henkin, of Georgetown University, whom we have referred to earlier, finds that zinc is good therapy for almost half the patients who come to him. One of his patients has the professional job of supplying bouquets to the Queen of England. Loss of her sense of smell could cost her a good job. Dr. Henkin is treating her with 100 milligrams of zinc daily.

Interestingly enough, **specialists in this field have found that sense of smell is closely related to sexual health**. Teenage girls who have delayed menstruation can be tested for sense of smell. If it is less acute than it should be, they may lack ova and need treatment. A teenage boy with small, inadequately functioning sex organs and inadequate sense of smell needs treatment, whereas if his sense of smell is normal, chances are he will develop normally in time.

A new and astonishing use for the trace mineral zinc has come to light in the treatment of an inherited disorder called *acrodermatitis enteropathica (AE)*, which is sometimes also called the Brandt Syndrome or the Danbolt-Close Syndrome. This hideous disorder usually attacks babies being weaned. It

36

used to be invariably fatal. Diaper rash on the baby develops into a thick red dermatitis and spreads to the baby's face, limbs, mouth, eyes, nose and other openings. Diarrhea soon appears, the baby loses its hair and gradually becomes weaker.

Dr. Edmund J. Moynahan of Guy's Hospital in London, England, an expert on this disease, has called it "a lethal, inherited human zinc deficiency disorder." He discovered this when he was treating three children in one family, all of whom suffered from intolerance to milk. They lacked the intestinal enzyme, lactase, which digests lactose or milk sugar. Probing for ways to feed such children, Dr. Moynahan did tests for all trace minerals and found that all three children were grossly deficient in zinc. Since this disease produces total baldness in the babies and, said Dr. Moynahan, since baldness is associated with lack of zinc, he gave the babies daily doses of zinc sulfate, which produced immediate, dramatic improvement.

All other victims of AE were immediately put on the same treatment. All of them improved. The original three babies are now thriving and have no further intolerance to milk. Dr. Moynahan gives up to 150 milligrams of zinc a day, in divided doses, since the children are suffereing from diarrhea. They lose considerable amounts of the trace mineral, as well as other nutrients.

The London specialist believes that the children have inherited protein deficiency in a certain digestive substance which results in tying up whatever zinc is in their intestines, making it unavailable for all the functions which this trace mineral performs in the human body. By giving plenty of zinc, the doctor is able to supply so much zinc in the intestine that it cannot all be made unavailable. Fortunately the disease is rare.

A physician wrote to the *Journal of the American Medical Association*, February 27, 1978, describing the case of a supposedly healthy young woman who gave birth to a healthy baby and, the next day, lost her sense of taste and smell. Her doctor thought this might have been caused by a viral illness a month before delivery. The doctor did X-rays and could find nothing wrong. The young woman began to take vitamin and mineral supplements. Her sense of taste and smell returned. Could these supplements have had anything to do with it, asked the physician?

Yes, they could, said the specialist from Georgetown University Hospital who answered the question. **It seems that, during pregnancy, zinc and copper are transferred from the mother to the unborn baby.** So lack of these minerals may be one cause for the loss of appetite that occurs in early pregnancy, along with peculiar eating habits and cravings. (The expectant mother who wants a midnight snack of pickles and ice cream).

Doctors do not know how much of these minerals may be lost to the mother in individual pregnancies. It probably varies greatly. So loss of taste and smell after childbirth probably indicates just a great loss of these minerals, since it is well known that zinc deficiency causes loss of taste and smell. For this reason, says the Goergetown specialist, the official estimate of need for pregnant women is now set at 25 milligrams of zinc daily—10 milligrams higher than for nonpregnant women.

Interestingly enough, *Medical Tribune* reported in August, 1977 that a team of doctors at the University of Minnesota had achieved "dramatic cure" of **Crohn's Disease** in a patient who had very low levels of this trace mineral. They gave him the zinc by mouth. Crohn's Disease, as we report in another chapter, is an infection of part of the intestinal tract which, up to now, has been treated almost exclusively by surgery.

It seems quite possible that modern American diets, of which half the calories are derived from refined carbohydrates, are responsible for this condition. **Practically all the zinc has been removed from such foods as white sugar and white flour and all products made from them.** So it seems evident that we must have a great national deficiency in this trace mineral.

Some people eat too much, put on excess weight and can't seem to get rid of it because of their voracious appetites which make them crave food continuously. Other people—chiefly people with life-threatening diseases such as cancer—cannot bring themselves to eat because they lack appetite, so they waste away. Apparently the final disaster is caused by this lack of appetite, which leads them to eat less and less until they are eventually malnourished and unable to overcome their disease.

Something in food which could be manipulated to treat both these conditions has been discovered by a Georgetown University researcher. The "something" is zinc. The physician and biochemist who has done the most work on zinc metabolism has said that all of biochemistry may have to be rewritten, now that we are discovering the great potential of zinc.

**Scientists have known for some time that lack of zinc in the body causes lack of appetite.** Well, then, reasoned Dr. Robert I. Henkin of Georgetown, who is both an M.D. and a Ph.D., why not try depleting the body stores of zinc to see if we can cut the appetites of overweight people so that they will voluntarily stop eating too much.

Dr. Henkin, whom we mentioned earlier in this chapter, had noticed that patients given the amino acid (form of protein) called histidine for progressive systematic sclerosis lost their appetites. The sensations of smell and taste disappeared, then appetites fell to the point where these people could not be persuaded to eat. If they continued to take histidine, they developed serious mental disorders which also suggested that they lacked zinc. Testing their urine and blood confirmed the fact that these patients were indeed seriously depleted of this trace mineral.

Why not give histidine to a group of healthy volunteers, Dr. Henkin said to himself, and see if we can cut their appetites. He gave a group of healthy young college students eight grams of the amino acid histidine every day. And within four to six days they complained of lack of appetite. So unwilling were they to eat that they lost as much as two to three pounds in the week during which they took the amino acid. And these were men who had never missed a meal in their lives, so healthy had their appetites been up to this time.

They had the least trouble eating breakfast. By dinner time they could hardly be persuaded to finish the meal. Since many overweight and obese people tend to skip breakfast, then eat ravenously at dinner and after dinner, a pill which would reverse this trend might be helpful indeed. All of us need to start the day with breakfast. But most of us tend to eat too much at dinner and after dinner when we are not going to engage in anything strenuous which will help to use up the added calories.

**How does histidine destroy appetite?** Apparently it depletes the body of zinc, very rapidly, through the urine. Giving small doses of the amino acid causes small losses of zinc. Larger doses bring about the larger losses. Every volunteer to whom Dr. Henkin gave histidine lost his appetite. The larger the doses of the amino acid, the more rapidly the appetite was lost.

But it's not a very good idea to lose one's appetite entirely, since dire troubles with malnutrition can result. So Dr. Henkin gave zinc to his volunteers and brought back their appetite miraculously, even though he was still giving them histidine. His research group has found and isolated a zinc-bearing protein in saliva called *gustin* which they believe is essential for maintaining proper health of the taste-buds which help to control appetite. However, says *Medical Tribune* for May 18, 1977, "The precise mechanism by which the desire to eat is suppressed is a combination of both gustatory and systemic factors, Dr. Henkin asserted, but it is not yet known in complete detail."

Now what about people at the other end of this teeter-totter of appetite versus lack of appetite? Cancer patients have consistently lower blood levels of zinc than healthy people and patients with other diseases. Cancer patients also tend to have less ability to taste and enjoy the taste of food.

"Feeding these patients small doses of zinc sulfate successfully diminished their anorexia (lack of appetite) and eventually their taste acuity was restored to normal, encouraging them to eat more heartily," said *Medical Tribune*. Dr. Henkin suggested that the lack of zinc in cancer patients may result from the growing tumor collecting the zinc in their bodies and making it unavailable for performing its many functions.

**So loss of appetite may become a vicious cycle.** The tumor is already depleting the cancer victim's body of many nutrients which could help the individual to fight off the cancer. Lack of zinc and the subsequent destruction of appetite makes this situation even worse.

Dr. Henkin thinks he may be on the trail of something very helpful for aiding the person who wants to take off pounds but can't control a galloping appetite, as well as helping victims of wasting diseases whose lack of appetite may arise from lack of zinc.

There is no way whereby his research with histidine can be used at this stage of the game by unsupervised individuals. As we saw above, this amino acid, given in large amounts by itself, can produce dangerous symptoms, partly because of depletion of zinc. So any further experimentation along these lines would have to be carefully controlled.

"Laboratory rats fed diets deficient in zinc ate only about one-third of the food eaten by rats getting enough zinc—apparently from lack of appetite caused by the lack of zinc," *Medical Tribune* reported.

**One cannot help but fear that many aspects of human health could easily be disrupted by depleting the body of most of its zinc.** This trace mineral is active—that means essential—to many, many human body enzymes. The list is increasing every year. This means that, in zinc deficiency, not one of these enzymes can function.

Such stories as these demonstrate the great complexity of the body's workings. They also show that it is impossible to state that this or that trace mineral can be used to alleviate this or that minor symptom. In the case of the AE babies, skin, digestive tract, hair and ability to sustain normal weight were all disordered by lack of one trace mineral.

Does this mean that all skin disorders may be caused by lack of zinc? All baldness? All diarrhea? All cases of inability to gain weight? Of course not. Does it mean that everyone who cannot taste or smell lacks zinc? Not necessarily. But such stories as this point up the vital importance of nutrients which we may need in extremely small amounts, but which we absolutely must have to survive. Any method of agriculture or food processing which discards trace minerals should come under heavy suspicion. Somewhere, somebody is going to suffer from this lack.

Although all of us do not have the inherited conditions which make large amounts of such trace minerals necessary, as the above patients did, all of us may have inherited certain needs for various nutrients that are much larger than normal. In these cases supplying the inadequate nutrient in much larger amounts may bring about complete alleviation of symptoms.

The other lesson to be learned from these stories about zinc is the terrible chance we take by exposing ourselves to anything known to cause mutations. These are the disordered enzyme patterns which cause inherited disease. They can be brought about by exposure to radiation, or by exposure of one's mother to radiation while she is pregnant. They can be caused by various drugs and many other chemicals to which we are often exposed in our modern technological society. This is why, when you read about any substance which is believed to be "mutagenic"—that is, capable of causing mutations—shun that substance.

We know that zinc is concentrated in the prostate gland and the pancreas, among other organs, thus probably indicating its great importance for the health of these organs. We know that it is important for the health of collagen—the "gristle" that suffers when we get arthritic diseases. We know that it is needed for bone growth and lack of it will cause dwarfism. We know that all muscles, including the heart, need zinc for efficient functioning. The eye needs zinc. Who knows what a variety of eye disorders might be prevented just by getting enough zinc? Zinc is essential for healing wounds and preventing infections. The list of essential activities of this trace mineral is very long indeed. And there is a great deal of evidence that millions of us may be deficient in zinc because of past eating habits.

**Wholegrains are a good source of zinc.** White sugar contains no zinc, although the body enzymes involved in processing sugar must require this trace mineral for efficient functioning. Alcohol depletes zinc from the body cells.

We could go on and on with lists of present-day circumstances that contribute to a possible nationwide deficiency of zinc.

So it is apparent that we must approach any therapy which depletes our bodies of zinc with great caution. People who are overweight may also have diabetes which may make their requirements for zinc greater than those of the rest of us. People who are alcoholics may have much greater need for zinc than the rest of us. So trying to reduce their weight by destroying appetite may produce side effects worse than the overweight.

But, at the other end of the scale, there seems to be every reason why people who lack appetite, for whatever reason, should take zinc on their own every day, for this trace mineral appears to be relatively nontoxic, except in massive doses such as eight or ten *grams* a day. Old folks often become emaciated from lack of appetite. Perhaps a zinc supplement every day can improve their desire for food. Sick people confined to bed often have little desire to eat, but they need the nutriment to help them recover. Zinc supplements may be the answer.

And surely victims of extremely serious, debilitating diseases like cancer should be taking zinc to improve appetite in order to prevent the wasting of muscles which occurs when

43

not enough food is eaten because the sufferer says, "I just don't want anything. I can't taste it."

To reiterate, **zinc is most abundant in these foods:** meat, eggs, fish, poultry, all dairy products, fresh fruits and vegetables (chiefly those that are bright green or bright yellow), wholegrain cereals and breads and all seed and nut foods. Zinc is removed entirely when white flour and processed cereals are manufactured. It is never replaced. Our national deficiency in this mineral may arise from this fact alone, as well as zinc deficiencies in soil in which food is grown.

# CHAPTER 4

# The Prostate Gland

THE PROSTATE is a gland located at the base of the bladder. The gland sometimes enlarges and continues to enlarge as a man gets older. The cause of this enlargement is not known; it affects about 30 per cent of all men, married and single, by the time they reach 50 years of age; and there is no known relationship to other infections or to individual sex habits, reports *Medigraph Manual*, by Dr. George E. Paley and Herbert C. Rosenthal.

**Cancer of the prostate** accounts for one out of every 10 deaths from cancer in men, the manual says. As with most malignant diseases, the cause is not known.

**When the prostate becomes enlarged,** the patient urinates more frequently and feels the urge to urinate several times during the night. As the disease progresses and the prostate enlarges, the bladder becomes thin, weak and less efficient. There is difficulty starting the stream of urine, and waste products are retained in the blood, leading to generalized weakness. In the final stage, the manual reports, the patient cannot pass urine at all.

Cancer of the prostate can be undetected until it has spread to involve other organ systems and bones, because often there are no symptoms and no discomfort. There may be mild discomfort on urination or blood in the urine. If there is any pain, it is felt in the area just below the rectum.

Complications in prostatic enlargement develop when obstruction, due to enlargement, prevents the free passage of urine and stagnant urine collects, the encyclopedia continues. There is back pressure and infection involving the ureter, urinary tract and the kidneys. At a late stage of the illness, the patient may be in great pain because the bladder is so overstretched. Uremia, the condition of having too much urea in the blood, may be an added result of this condition. That is why it is recommended that men 50 and over should have a urine analysis and prostatic examination, along with a regular physical check-up, every six months.

**The male sex organs of mammals are extraordinarily high in zinc,** especially the prostate gland which is where the male sperm is stored. The sperm cells themselves are high in zinc. In humans, the male prostate gland contains more zinc than any other organ of the human body—except for certain parts of the eye—102 micrograms per gram—almost twice as much as the liver and kidney. All body muscles, including the heart, store zinc, indicating that this mineral is apparently very important to the operation of muscles and heart. The pancreas, lung, spleen, brain, testes and adrenal glands also contain appreciable amounts of zinc.

"Do oysters make you sexy? They contain zinc, without which the male gland cannot perform," report Dr. E. Cheraskin, Dr. W. M. Ringsdorf, Jr. and Arline Brecher in *Psychodietetics.* "Dr. James Leathem of the Bureau of Biological Research at Rutgers University has demonstrated that zinc and manganese are both indispensable to a healthy prostate and may help a degenerating gland to recover."

Writing in *Nutrigenetics,* Dr. R. O. Brennan and William C. Mulligan report that men often cannot produce sperm when they suffer from severe vitamin B-complex or vitamin E deficiencies and that the supplementation of these vitamins has been found to correct this disorder. In addition, they add, **the prostate gland is partially responsible for the secretion of seminal fluid, and a deficiency in zinc can cause this gland to malfunction.** Only half the normal

46

amount of zinc is found in the pancreas of diabetics, who are often sterile, they say.

Today, prostate gland troubles are almost universal among older men in our part of the world, and they are becoming increasingly common among younger and younger men. **Doesn't it seem possible that deficiency in zinc, either partial or extensive, may have something to do with this epidemic of disorders?**

Most American men are brought up on diets in which processed cereals, white bread and other foods made of white flour are staples. Since the zinc has been removed from all these and never replaced, could not this single factor explain why prostate gland problems are so prevalent in Western society and almost unknown among more "backward" people who are still eating unprocessed, unrefined cereal products?

In *New Age Nutrition*, by Richard J. Turchetti and Joseph M. Morella, we learn that 23 of the 50 states are devoid of zinc, and "the food grown in those soils is subsequently devoid."

Therefore, is it not possible that the swelling of the prostate gland is reacting the same way the thyroid gland does when it desperately needs iodine? But the prostate is in need of zinc, perhaps.

For some reason, not yet understood, **the levels of zinc in blood vary from one geographical region of the United States to another.** Could this be, as we just reported, because soil or drinking water in some parts of the country lack zinc? No final explanation has been made, but, in view of the fact that the amount of zinc available for human beings is "marginal," perhaps we should become concerned about this discrepancy. And, as we report in this book, a zinc deficiency may be a contributing factor to a long list of disorders now affecting many Americans of all ages and all walks of life.

"No part of the human anatomy contains more zinc than the prostate gland," says Robert Rodale in *The Best Health Ideas I Know*. "Actually a healthy man has several times more zinc in his prostate than he does in most other soft tissues of

47

his body." He goes on to tell us that large doses of zinc have improved prostate health in experiments conducted in Chicago. Scientists found that seven per cent of all men they tested have low levels of zinc in their semen and 30 per cent are borderline. They have given zinc to men suffering from chronic prostatitis (not caused by bacterial infection), have gotten good results in 70 per cent of these.

If zinc is obviously essential and involved in so much that goes on inside us, why are not doctors more concerned about it and why do they not give their patients zinc supplements? Rodale lists some reasons. Doctors are notoriously uninterested in nutrition. The typical American diet, which is causing zinc deficiency, is thought by most doctors to be the best possible diet. Then, too, giving trace minerals is a way to prevent disease, rather than healing a condition already present. Doctors are looking for quick cures for acute diseases, mostly. They seldom express much interest in preventing disease, except by vaccination.

Finally, writing in *Let's Eat Right to Keep Fit*, Adelle Davis says that vitamin A and vitamin C may be useful in keeping the prostate healthy.

# CHAPTER 5

# Arthritis

THE WORD "COLLAGEN" is defined in the medical dictionary as "the albuminoid (protein) substance of the white fibers of connective tissues, cartilage and bone. It is converted into gelatin by boiling." So it includes all those body tissues which "hold us together." In his book, *The Healing Factor, Vitamin C Against Disease*, Dr. Irwin Stone **defines collagen as "The main structural protein of the body, comprising about one-third of the protein content of the body.** This is the cementing substance that holds the tissues and organs intact, forms and maintains the integrity of the vein and artery walls, lends strength and flexibility to the bones, and is the main component of scar tissue and healing wounds. The body cannot produce collagen without ascorbic acid (vitamin C). The most distressing symptoms of scurvy are caused by defective or absent collagen." Scurvy is the disease of vitamin C deficiency.

The medical literature has long discussed **the relationship of vitamin C to collagen,** stressing the fact that this relationship is what makes vitamin C so valuable in the performance of so many functions in the body. After all, if vitamin C, in given amounts, is necessary for producing and maintaining the health of everything in our bodies that connects everything to everything else, it surely must be involved with just about everything that goes well or goes

49

wrong in our bodies.

The *Stein and Day International Medical Encyclopedia* describes the collagen diseases as a group of diseases characterized by changes in the collagen fibers which help to make up the fibrous supporting tissues of the body. These diseases may affect the brain, heart, joints and subcutaneous connective tissue, and are polyarteritis nodosa, rheumatic fever, lupus erythematosus, scleroderma, dermatomyositis and rheumatoid arthritis.

All the various conditions that are grouped under the head of arthritic diseases are also called "collagen" diseases, since, as we report, it is the connective tissue which is disordered or inadequate in these conditions.

Recently, we came upon some information on zinc which seems to indicate that it, too, is extremely important in regard to collagen—its formation and its good health. A new book, volume one of a series called *Trace Elements in Human Health and Disease*, deals with the trace minerals zinc and copper.

In a chapter on collagen, the authors, Felix Fernandez-Madrid, Ananda S. Prasad and Donald Oberleas, discuss **the effects of zinc deficiency on the creation of this connective tissue called collagen.** They describe experiments which show that the effect of zinc deficiency on the manufacture of collagen is a generalized effect on the manufacture of protein and on the workings of nucleic acid, rather than a direct effect on the manufacture of collagen.

For example, deficiency in zinc produced (in laboratory rats) a great decrease in the amount of three amino acids or forms of protein in the skin of the animal. These three (glycine, proline and lysine) are especially important ingredients of collagen. So lack of zinc, resulting in a lack of these three kinds of protein, could be very important, indeed, to the health of the skin, as well as other tissues.

This suggests, does it not, that anybody with any kind of skin condition might notice improvement by simply increasing the zinc in his or her diet. It also suggests that the almost universal complaint of skin ailments these days—

especially among teenagers—may have a lot to do with lack of zinc in their diets.

It is well known that **wounds do not heal as quickly as they should in living things which are deficient in zinc**. Giving zinc improves this situation. Zinc accumulates at the site of an injury, which seems to demonstrate that the body sends it there to help in healing, as white blood corpuscles are sent immediately to a wounded area to help in healing.

It is well known, too, that children who are deficient in zinc do not grow as they should, which seems to indicate that they do not have enough of this trace mineral to help them make the protein necessary for all the collagen in bones, cartilage and other connective tissues. "In general," say the three authors, "studies of a variety of connective tissues in the zinc-deficient state have shown conclusively a significant reduction in the total collagen in the zinc deficient state."

People deficient in zinc also show abnormalities in the way the body uses those important substances DNA and RNA, which are the elements in cells which regulate heredity. As each cell divides, it takes with it part of the original RNA and DNA, so that the next cell will be normal. These substances are also made of protein. If zinc is essential in making protein, then it seems likely that the creation and operation of RNA and DNA (also called nucleic acids) would suffer. For the record, DNA stand for deoxyribonucleic acid and RNA is the abbreviation for ribonucleic acid.

It seems, indeed, from several reports, that **zinc participates in the manufacture of nucleic acids**—RNA and DNA. In animals made deficient in zinc the body manufacture of DNA was impaired. DNA is necessary for cells to divide. Cells must divide in order to create collagen. So the breakdown of this entire process seems to be implicated in reduction of collagen which develops in animals deficient in zinc.

There is every reason to believe that the same thing happens in human beings. **Lack of zinc causes disordered collagen health and inefficient repair.** Now think, for a

51

MOVIES TO GO
808 WEST SUNSET
SPRINGDALE, AR 72764
756-8660

moment, of the many parts of the body which are likely to suffer from such a circumstance. Most of all, think of the millions of Americans imprisoned in a "collagen disease" of one kind or another.

**At present some three and a half million Americans suffer from some form of collagen disease.** According to the Arthritis Foundation, about 97 per cent of all Americans over the age of 60 have collagen diseases of greater or less severity. In other words, these diseases are almost universal among older folks.

A physician from Seattle, Washington has been treating arthritic patients with a daily zinc supplement given with each of the day's three meals. He reports that results have been encouraging.

*The Lancet* for September 11, 1976 contains an article by Dr. Peter A. Sinkin, of Seattle, describing a trial of zinc supplements for 24 of his patients with "refractory" **rheumatoid arthritis**. This means arthritis which did not respond to any other method of treatment. It was a double-blind experiment.

That is, the volunteers were divided into two groups. One group was given their regular medication plus zinc supplements for 12 weeks. The others got their regular medication plus a tablet which contained nothing. Not until the end of the trial when all observations on joint swelling, morning stiffness, walking time had been made and the patients themselves described how they felt was the code broken so that patients and doctor alike knew which had received the zinc supplement.

The doctor examined all patients at the beginning of the trial, recording scores for swelling of joints, tenderness in each joint, along with scores for the length of time morning stiffness persisted, and grip strength, as well as ability to walk 50 feet in less than 30 seconds. Patients reported any symptoms they had of such things as discomfort, nausea, vomiting, change in appetite or bowel habits and so forth. X-rays of all hands were taken at the beginning of the test, again at 24 weeks.

Says Dr. Sinkin, **"patients taking zinc sulfate fared better in all clinical parameters than did patients receiving placebo** (the nothing pill)." In every area investigated—swelling, pain, stiffness, walking time, the ones getting the mineral improved while those who did not take the zinc showed little or no improvement. The only test in which the zinc did not seem to bring much improvement was the grip strength. An early improvement in the test group was not sustained, says Dr. Sinkin.

After the original test, **all the volunteers in both groups were given zinc supplements and all reported improvement**. There were few side effects. Headache, rash, change in appetite, abdominal pain or discomfort and diarrhea were all reported oftener in the group *not* getting the zinc supplement than in the group which got it. All side effects were mild, however.

The zinc sulfate was taken after meals to prevent any difficulties with nausea. Dr. Sinkin suggests that it would be preferable to take a zinc supplement which would be better absorbed and would be taken without any digestive irritation. "From our experience," he says, "and that of others, virtually all patients can tolerate oral zinc sulfate for three to six months. Possible toxic effects of prolonged use must still be carefully sought." He believes that much additional work must be done to confirm his observations and to determine what part zinc plays in the health of joints.

**Officially, adults are supposed to get 15 milligrams of zinc daily, while children need less, depending on age.** Meat, liver, fish and shellfish are the best sources of zinc, cereals which have not been refined or processed come next. But white flour products and supermarket cereals contain no zinc. Liver, beef, peas, corn, egg yolk, brewers yeast, carrots, dairy products and brown rice are good sources. It's easy to see why many Americans might be short on zinc since over half their diet may consist of foods made from white flour and white sugar, from which all the zinc has been removed and never returned. Isn't it possible that just this one shortage area may be responsible for our modern plague of arthritic diseases, which torment and deform millions of Americans?

If you suffer from arthritis, there is no reason to delay in taking a zinc supplement. There is no need to take the sulfate form which Dr. Sinkin gave to his patients. Your health food store has zinc supplements which are readily absorbed. These come in various potencies, marked on the label. An especially valuable kind is the "chelated" product, meaning that the mineral has been associated with amino acids which "chelate" it into a form which is much more readily absorbed by the body.

Dr. Sinkin's patients were taking large amounts of zinc three times daily. Much more zinc would undoubtedly be absorbed from a chelated product than from the zinc sulfate tablet which Dr. Sinkin gave. So there seems to be no need to take as much as he was giving patients in his experiment. Eat lots of those foods which contain plenty of zinc and avoid those from which it has been removed. And take a zinc supplement. It is a perfectly natural substance which cannot harm you. There is now a recommended daily dietary allowance for zinc, meaning that we must have specific amounts of this necessary mineral every day.

# CHAPTER 6

# Zinc for Surgical Patients

CALAMINE LOTION, prescribed by doctors for many skin conditions, is simply zinc oxide with a small amount of ferric (iron) oxide. Calamine lotion was used to treat skin and eyes before 1500 B.C. Since that time it has been made up into salves, powders, ointments which have been very effective against wounds and sores of many kinds, as well as skin eruptions from unknown causes and poisons like poison ivy.

Reading medical journals, one wonders why there are so many accounts of non-healing ulcers and surgical wounds that do not heal, when zinc preparations might be used to treat these conditions.

In his massive book, *Trace Elements in Human Health and Disease, Volume I, Zinc and Copper*, Ananda S. Prasad devotes one entire chapter to the subject of zinc in relation to the surgical patient. He tells us that **"the effectiveness of zinc therapy for healing and repair is dependent on various factors that modify the metabolism of this element and healing in general.** Some of these factors include age, sex, infection, nutritional status, medications, including steroids, interaction with other elements and vitamins, and use of radiation for therapeutic purposes."

He also said that, "**The metabolic roles of zinc appear to be so numerous that biochemistry may have to be rewritten around this one element.**"

Workers in animal nutrition have rediscovered zinc as a healing factor and for the past 24 years or so have been turning up astonishing accounts of animals that healed more rapidly when they were given zinc. In practically every case the addition of zinc to the diet aided in healing. Chelated zinc supplements brought faster healing than plain zinc. Chelating a trace mineral with an amino acid causes it to be absorbed more quickly and more successfully than otherwise.

The skin of domestic animals is especially sensitive to zinc deficiency, says Prasad. Cattle develop a dry scaly skin, thickening of the skin around the nostrils, horny overgrowth of mucus membranes, baldness, red, scabby, wrinkled skin on the scrotum or tender, easily injured and often raw and bleeding skin. Wounds heal slowly if at all. The same conditions prevail in chickens and swine who are deficient in zinc.

In 1955, two scientists discovered that deficiency in zinc is a predisposing factor in a disease called *parakeratosis* in swine. Since then there has been an "explosion" in zinc research, says Prasad. Food animals are expensive, so there is always plenty of money available for research on their diseases. If any information spills over that might be valuable for human beings, so much the better. And this has been true to some extent with zinc.

During the past 10 years, says Prasad, evidence has accumulated that **hospital patients are frequently deficient in zinc**. Poor healing results. And giving zinc in supplements helps wounds to heal. The reasons for zinc deficiency in surgical and other hospital patients seem to arise from a combination of poor eating habits before coming to the hospital, inadequate intravenous feeding (if that has been used), metabolic diseases which include, it seems to us, just about every condition of modern life and especially those diseases which send so many people to the hospital for operations. Then, too, it seems that zinc is lost in large amounts in both urine and feces, especially in the surgical patient.

**Here are some of the conditions in which patients have been found to be deficient in zinc:** alcoholism, hardening of the arteries, chronic skin ulcers, cirrhosis of the liver, Down's Syndrome, dwarfism, lung cancer, malnutrition, pregnancy, pulmonary infection (including TB) and uremia. We assume that many more conditions could be listed here if any efforts had been made to look for zinc deficiency.

Prasad said flatly, **"Today all patients presenting with chronic disease should be studied for zinc deficiency."** It develops very rapidly in burn patients. Leg ulcers associated with many varying conditions are common in zinc-deficient patients. Venous ulcers the same. The skin of people with poorly healing ulcers has been found to be deficient in zinc. Fortunately, says Prasad, these conditions respond well to zinc supplements given orally.

For example, an elderly woman and an infant had advanced zinc deficiency. Both had thin, glazed, fragile skin. The woman's skin broke down repeatedly and would not heal. Both patients returned to normal with zinc therapy. Bed sores can and will respond to the treatment with zinc.

Today many people are on therapy with the steroid drugs—derivatives of cortisone, for example. Long-term use of these tends to create zinc deficiency, hence poor healing of wounds. But other glandular substances may also disorder the body's use of zinc—the sex hormones, for example. What about The Pill, which is one of these? Several researchers have reported disorders of the body's use of zinc as well as zinc deficiency in women on The Pill.

We know, says Prasad, that anesthesia causes the blood levels of zinc to fall in the surgical patient. We know, too, that the surgical wound itself causes zinc levels to fall. So after the operation the surgical patient is left dangerously low in the one trace mineral that is essential for healing!

57

Why do not doctors realize this and give their surgical patients ample amounts of zinc supplements before and after surgery? We don't know why except that surgeons in general appear to be most concerned with the actual mechanics of the operation, and not with what happens to the patient before or after the operation. And the GP, who takes care of the patient after the surgery, doesn't seem to have much information on the possible nutritional losses that may occur directly as a result of the surgery.

We know that vitamin C is rapidly destroyed in the body of someone who has been operated on. Loss of vitamin K due to antibiotics given the patient before the operation is likely to be responsible for hemorrhages. Now we read that loss of zinc, too, may further endanger the course of recuperation after surgery.

Why should so many people show up in the hospital deficient in zinc to begin with? **Well, when cereals are refined to make white flour, and sugar cane is refined to make white sugar, all the zinc in each of these foods is removed.** So an entire nation of people who have been brought up on a diet more than half of which is made up of these two foods, is bound to be deficient in zinc, as well as all the other trace minerals destroyed in the refining process.

**Zinc is important to the health of the pancreas and the prostate gland in men.** Today, we are in the midst of an epidemic of pancreatic disorders (diabetes, low blood sugar, pancreatitis, pancreatic cancer) and prostate disorders which affect practically all men past middle age. Doesn't it seem possible that the almost complete removal of this one trace mineral, zinc, from our meals over the past 50 years or so may be largely responsible for these epidemics?

When a circulatory disease develops as a result of diabetes, surgery is frequently resorted to and the zinc-deficient patient comes out of the surgery even more deficient in the one element—zinc—which is essential for quick healing of the surgical wound! When a victim of prostate gland disease is operated on to remove the gland or to remove any tumors

58

that may have affected it, this zinc-deficient individual comes out of the operation even more deficient in this one element which is so essential for healing.

If these surgical patients return home to the same diets that created their need for surgery, they are bound to invite more and more complications as a result of worsening zinc deficiency, as well as deficiency in all those other trace minerals missing from white sugar and white flour, the B vitamins, etc.

**Can you remedy zinc deficiency by diet?** Of course you can. Just eliminate from your meals and snacks any and every food that contains sugar. Make your desserts a piece of fresh fruit, or some cheese and nuts, or popcorn, dried fruit, nuts or sugar-free snacks from the health food store. Make your cereals only the wholegrain ones, your breads only the real wholegrain ones. And take zinc supplements. They are available at your health food store. The chelated products are absorbed and used most effectively.

"Whenever there is tissue injury—from a severe accident, extensive operation or destruction of tissue by cancer or infection—alterations in the constituents of blood occur which favor blood clotting," says Dr. Alton Ochsner, Professor Emeritus of Surgery at Tulane University School of Medicine. The body at all times responds to this kind of injury by activating its mechanism for clotting blood so that we do not bleed to death.

So there is special danger from blood clots after major surgery. In addition, the patient is lying quietly in bed. Blood in the veins of the legs is normally moved along and returned to the heart chiefly by the action of leg muscles, especially calf muscles. So lying quietly in bed without moving these muscles invites stagnation of the blood in veins. This is the main reason why it is now usual procedure to get surgery patients on their feet as soon as possible after operations. But we should also keep this consideration in mind when other illnesses or accidents confine us to bed. Blood clots are quite likely to form in our legs and move without warning into heart and/or lungs, possibly causing death.

This condition is called phlebothrombosis. In 1948, a colleague of Dr. Ochsner's at Tulane discovered that vitamin E, in the presence of calcium, decreased the incidence of clotting in a testtube. The vitamin was then given to patients at the medical center there who had severe injuries, hence were candidates for fatal blood clots. The incidence of these clots decreased decidedly. "It is for that reason that I have been using alpha tocopherol (vitamin E) since that time and have become very enthusiastic about its use in the prevention of intravenous clotting," says Dr. Ochsner in *Executive Health*, Vol. X, No. 5, 1974.

He also uses vitamin E for long-term prevention of unhealthful clotting, since it carries none of the hazards of anti-coagulant drugs. These drugs, unless monitored constantly, can cause hemorrhaging which can, of course, also be fatal. Vitamin E does not produce this tendency in the blood.

Dr. Ochsner tells the story of a 43-year-old man whose blood indicated a tendency to clot. He developed pain in his right knee and hip along with a condition usually caused by a clot in the blood vessel of that region. The doctors decided to operate and replace the hip—an extremely serious operation. They gave the patient 200 units of vitamin E three times daily, then 100 units of vitamin E three times daily until they brought the clotting condition of his blood back to normal. Then they performed the operation, giving him more vitamin E after the operation. He recovered with no complications. **"I am convinced that its (vitamin E) use is extremely beneficial and at times life-saving,"** says Dr. Ochsner.

He points out in the same article that **vitamin C is also helpful before and after surgery**. It, too, helps to control clotting in veins. He refers to a British experiment in which 30 of 63 surgery patients were given one gram (1,000 milligrams) of vitamin C daily before and after their operations. These were people up to the age of 84 undergoing very serious operations. Of the 30 patients given the vitamin C, only 10 had positive tests for clots, none had trouble in both legs and one had pulmonary clot. Of the 33 patients who were not given vitamin C, 20 (60 per cent) had positive tests for clots, in 10 of them both legs were involved and 12 patients had pulmonary clots. A most convincing bit of evidence.

Vitamin C, says Dr. Ochsner, is absolutely necessary for proper healing of wounds and by wounds he also means infections of any kind, for all infections cause injury to tissues. He describes a patient with extensive cancer of the jaw. Dr. Ochsner removed the malignant tissue but the wound did not heal. Instead it became infected. This was before the days of antibiotics. He gave the patient large doses of vitamin C, which controlled the infection and healed the wound.

Dr. Ochsner points out that **our supply of vitamin C appears to decrease as we grow older,** making it important that we take pains to get enough of it not only to provide for this shortage, but also to strengthen us against the stresses of old age—the falls, the infections, the broken bones, the possible surgery and so on. He quotes two scientists who say they found a gross and often complete deficiency of vitamin C in the arteries of apparently well-nourished hospital patients who died. We mention these vitamin deficiencies which occur after surgery to indicate to you the grave consequences of surgery on nutritional welfare. And loss of zinc is one of these.

**Vitamin C also appears to be involved in the cholesterol controversy.** Hardening of the arteries seems to be associated with increased cholesterol in the blood. A number of experiments have shown that some people with high blood cholesterol levels show a reduction in these levels when they are given large doses of vitamin C over a long period of time.

In one experiment, 92 of 106 patients with hardening of the arteries (a condition related to zinc deficiency, remember) had significant decreases in blood cholesterol—up to 30 per cent. In another test, diet was found to have no effect on cholesterol levels, but only one-half gram (500 milligrams) of vitamin C twice a day brought about a decrease of 30 to 40 per cent. When the vitamin C was discontinued, cholesterol levels rose once again.

Still another researcher found that another kind of blood fat, the triglycerides, were reduced when vitamin C was given in large amounts—2 to 3 grams of vitamin C daily for up to three years. Are such doses harmless? Yes, says Dr. Ochsner, who has prescribed such doses himself for many patients. Some scientists have questioned the possibility that vitamin C may be changed in the body into harmful amounts of another substance, oxalic acid. Except in very unusual individuals, says Dr. Ochsner, there appears to be no evidence that this will happen with doses of four grams or less daily— that is 4,000 milligrams or less.

Several other significant aspects of our essential need for vitamin C are these. Levels of vitamin C in blood are lower as we grow older and also in the white blood cells. The white blood cells are the body's defense against infections. When infection strikes, they are summoned to the spot and fight the germs. Apparently vitamin C is a vital part of this battle.

**Smoking destroys vitamin C.** Blood levels of the vitamin are consistently lower in people who smoke—so much so that some scientists speak of smokers living in an eternal state of subclinical scurvy. Men have lower blood levels of vitamin C than women, suggesting that they need to get more in their meals and supplements than women need. People who are at risk as heart attack victims have lower vitamin C levels than those who have no circulatory complications.

And one of the objections to the use of The Pill—the oral contraceptive—is that in some women it produces slight clotting in veins which can lead eventually to quite serious conditions such as strokes. One group of investigators has found that women taking The Pill have only about one-half as much vitamin C in their blood as those not taking The Pill. Dr. Ochsner asks if this might not be the reason for the increased incidence of blood clotting in women on this medication.

We all know the helpfulness of vitamin E and vitamin C in preserving good health and protecting us from many stresses and illnesses. Now we have the testimony of one of the world's great surgeons—Dr. Ochsner, who is one of the first physicians to appreciate the dangers of smoking. He wrote a book many years ago presenting evidence incriminating smoking in lung cancer and warning against the use of cigarettes.

Foods richest in vitamin E (many of which are also good sources of zinc) are wholegrain cereals and breads, wheat germ and bran, eggs, nuts, seeds and deep green leafy vegetables, as well as wheat germ oil and other salad oils made from other seeds and vegetables.

Foods richest in vitamin C are fresh raw fruits and vegetables (again many of these are also good sources of zinc), chiefly citrus fruits, strawberries, all vegetables in the cabbage family, green peppers, rose hips and liver.

# CHAPTER 7

# Zinc for Successful Childbirth

THE EVIDENCE OF the necessity of the trace mineral zinc for successful childbirth and mentally sound offspring is secured by experiments with laboratory rats. The evidence is so dramatic, so shocking that we are appalled to consider the consequences of zinc deficiency in terms of human life and health—that is, human mothers and babies.

As long ago as 1968, J. Apgard demonstrated that female rats on a diet deficient in zinc have a long and difficult pregnancy and show peculiar behavior during childbirth and after the offspring are born. The mother rats fail to make nests for their babies, fail to nurse them successfully, fail to gather the blind babies together for nursing. Feeding the pregnant rats plenty of zinc beginning on the 19th day of pregnancy completely prevented all these disastrous events.

In a more recent investigation, female rats were fed a diet low in zinc throughout their pregnancy. More than one-third of the zinc-deficient rats became ill during pregnancy—a "depressed state similar in some respects to physiological shock—cold to touch and in a state of torpor." The actual delivery of offspring was prolonged and some rats died during the delivery process. There was excess loss of blood. And 42 per cent of all the offspring were dead at birth. Most surprising of all was the behavior of the zinc-deficient mother rats toward those pups which lived. It is well known that healthy animal mothers will defend their babies against predators many times their own size and will fuss obsessively with the care of their young. They nurse them carefully, lick them clean, gather them together so that all offspring remain warm and snug in the nest.

The zinc-deficient mother rats in this experiment paid little or no attention to their pups. They did not nurse them or make efforts to keep them all together in the nest. None of the pups survived.

In the report on this experiment, published in *Nutrition Reviews*, October, 1977, there are an additional two paragraphs which are startling. **"It has been reported,"** says the publication, **"that ingestion of high doses of aspirin during pregnancy will cause a condition which resembles that described for zinc deficiency."**

One researcher reported that aspirin given to the mother rats daily from the 19th day up to delivery brought about 71 per cent mortality among the pregnant rats and 45 per cent mortality among those pups which were born. Giving a somewhat lower dose of aspirin brought about 59 per cent mortality among the mother rats and 40 per cent mortality to their pups. We wish these researchers had gone farther into this matter. Did the aspirin destroy even the small store of zinc these animals had, or was the action of the aspirin due only to some other aspect of this drug?

It is clear, says *Nutrition Reviews*, that an intake of zinc which is enough to prevent the birth of grossly abnormal offspring is not enough to permit the many bodily adjustments that must be made to carry a successful pregnancy through to a successful delivery.

What does such a finding portend to the young women of today who were brought up on diets deficient in zinc and perhaps frequent doses of aspirin? Diets in which sugar and other refined carbohydrates predominate are almost bound to be deficient in zinc. How will these women manage to carry through a successful pregnancy, and in what condition are their babies likely to be born? **Doesn't it seem possible that maternal neglect of human children may be linked to deficiencies in trace minerals?** If an animal mother stops caring for her newborn offspring because of zinc deficiency, couldn't the same thing happen to a human mother?

And what of the children that survive this kind of pregnancy? Are they handicapped in any way? According to a chapter in a massive book edited by Ananda S. Prasad, *Trace Elements in Human Health and Disease, Volume 1, Zinc and Copper*, there is good reason to believe that zinc deficiency in the mother impairs the emotional and mental well-being of the children born to her.

In a chapter in the book titled "Psychobiological Changes in Zinc Deficiency," Prasad and his colleagues describe experiments in which 30-day-old rats were placed on diets deficient in zinc and adequate in all other nutrients. They became lethargic. They did not gain weight as they should have. They could not successfully perform certain tests of learning ability. Their "emotionality" was increased. In rats this means they did not move around normally, they were not as active as healthy rats are.

Then the scientists tested the effects of only a mild zinc deficiency in pregnant rats to see what the effect on pregnancy and the learning ability of any offspring might be. The results were dramatic.

The mother rats suffered "severe disturbances" in delivery of their young and in behavior toward them after they were born. Up until one week before delivery the pregnant rats on a diet that was only slightly deficient in zinc looked and acted much like a second group of pregnant animals on a diet which included plenty of zinc. In spite of this, almost one-fourth of the zinc deficient rats died just before they could deliver their young. After delivery another one-fourth died.

Those mothers that lived had great difficulty in bringing their offspring into the world. There was premature labor and delivery. The mother rats failed to attend to the time-honored duties of a new mother in the animal world—biting off the umbilical cord, cleaning and nursing the pups, bringing back to the nest any wandering pups. In fact the deficient rats had failed to build any nests at all. Many of the pups which died simply starved to death due to maternal neglect.

Most significant of all, 71 per cent of all the pups born to the mothers who were only slightly deficient in zinc were dead by the time of weaning. The scientists planned to give all families of rats behavior tests, but so few progeny remained alive from the zinc-deficient offspring that this was all but impossible.

However, in tests that were done on adult progeny of zinc-deficient mother, "offspring manifested impaired behavior on all (test) measures," say the scientists. **The results clearly indicate the importance of dietary zinc for normal behavior development at different levels of maturation and point to the need for further research on the competition of mother and fetus for marginal nutrient levels."**

The pregnant mothers were slightly deficient in zinc; the unborn babies—in competition with their mothers for this small store of zinc were also not getting nearly enough. And this is on a diet only slightly deficient in zinc, with all other minerals, trace minerals and vitamins supplied in abundance!

*The Lancet,* a British medical journal, carries news in its March 22, 1975 issue of **a possible relationship between deficiency in zinc and the birth of babies with crippling deficiencies.** Multiple bone deformities and dwarfism resulted when three mothers who were deficient in zinc had children.

It goes without saying that any human mother deficient in zinc would probably also be deficient in many other trace minerals as well as vitamins, unless she was taking supplements to supply them. The same kind of human diet which lacks zinc lacks other minerals as well, and lacks vitamins, especially the B vitamins.

How does it happen that diets for laboratory animals are designed to nourish these creatures completely in good health, except when scientists remove one or another nutrient to test the results? This is how it happens. Laboratory animals have no choice when it comes to food. They can eat only what is put before them. They don't ever shop in supermarkets where more than three-fourths of all the food attractively arrayed on the shelves is grossly deficient in many minerals. They aren't plied with sugary goodies on every hand, every day, everywhere they go. They do not drink, imbibe coffee or tea, or smoke.

Most important, their diets consist in large part of whole grains and seeds, with all nutrients intact, designed to provide exactly what is needed by that kind of animal at every stage of life. Their air and water are protected from pollutants of any kind. But, in spite of this care, just removing one trace mineral—zinc—from their meals can result in the disastrous consequences described above.

It makes one wonder how any of our young folks manage to survive and produce healthy children, for practically all of them have been brought up on meals and snacks so denuded of trace minerals that no laboratory animal could survive on them. And every sugary snack crowds out of this deficient diet some highly nutritious food that could have been eaten instead.

We hope that all obstetricians in the country and all organizations like the Society for the Protection of the Unborn Through Nutrition (SPUN) will learn of these hazards of zinc deficiency and spread the news far and wide. Our prospective mothers must be told and told again of the hazards to them and their children of diets deficient in any essential nutrient.

SPUN has excellent literature available on diet and supplements for the pregnant woman. Their address is: 17 North Wabash Avenue, Suite 603, Chicago, Illinois 60602.

In addition, Dr. Tom Brewer, president of SPUN, and his wife, Gail Sforza Brewer, have written an excellent book which discloses the terrible shortcomings in nutrition instructions being given to many pregnant women by their obstetricians. The Brewers also recommend a sound diet for the pregnant woman which will nourish not only her, but her unborn child, so that both will come through childbirth in good health.

The title of the book is *What Every Pregnant Woman Should Know. The Truth About Diets and Drugs in Pregnancy*. It is published by Random House in New York at $8.95.

# CHAPTER 8

# Acne

ACNE IS AN inflammation of the sebaceous (oil) glands just beneath the surface of the skin, causing pimples, blackheads, whiteheads, and, in extreme cases, infected cysts and scarred skin, according to *Reader's Digest Family Health Guide and Medical Encyclopedia*. Usually setting in at puberty, cases of acne may be brief and mild or chronic and severe, but, in one form or another, **it affects as many as 80 per cent of all teenagers**. Areas most frequently affected are the face, shoulders, chest and back.

"When adolescence begins, glandular activity increases," the encyclopedia adds. "The sebaceous glands secrete a greater amount of the oil, called sebum, that lubricates the skin. The excessive oil clogs the pores, causing them to dilate. This increases the likelihood of further infection.

"A mild case of acne will consist of no more than a few pimples and blackheads. If the sebum seeps under the surface of the skin instead of coming out to the surface, the surrounding tissue is irritated, and a cyst is formed. At this stage, your doctor should be consulted.... In severe acne, the skin of the upper torso and the face may be so seriously damaged that permanent scars will result. Prompt medical attention to early symptoms is, therefore, extremely important."

It is important that hands be kept away from the face;

squeezing pimples, blackheads or whiteheads with fingers must be avoided, and a special effort must be made to stop any unconscious picking at scabs. Washing the skin too vigorously or with a washcloth that is not absolutely clean can inflame a sensitive area, the encyclopedia states.

It adds that mild cases of acne (often called *acne vulgaris*) are usually brief and are likely to disappear if proper care is exercised with regard to diet and cleanliness. Chocolate, fatty and fried foods, cola and other sugar-rich drinks, cakes, pastry, pies and alcoholic drinks are some of the offending foods and beverages which should be avoided. And so should candy and other sugary treats.

**Certain chemical forms of vitamin A have been prescribed successfully by some American physicians in the treatment of acne. Now we read of several Swedish physicians who are using vitamin A and zinc to treat this unsightly skin disease.** According to *Archives of Dermatology* for January, 1977, Dr. G. Michaelsson and his colleagues gave zinc sulfate in very large doses (135 milligrams daily) alone and in combination with vitamin A. The results were compared with results from vitamin A alone and a pill containing nothing.

To decide just how effective each of these treatments was, the doctors counted the number of "pimples" at each visit. After four weeks, there was significant decrease in the number of such skin flaws in the group taking zinc. Those taking zinc and vitamin A got no better score. And the group taking the 'nothing" pill showed no change.

Most important of all, after 12 weeks of treatment the score on number of pimples had decreased from 100 per cent to 15 per cent. It seems to us that any teenager should welcome such a treatment with glee, and should persist in it until he or she completely clears those ugly pustules from the skin.

We have emphasized many times our conviction, bulkwarked by a great deal of significant research, that **acne is caused by the consumption of too much refined**

**carbohydrate**—that is, too much white sugar and white flour and foods made from them. No one seems to know just why this diet should cause acne, although teenagers in primitive countries who do not have access to these foods never suffer from this disease.

Now it seems that **lack of zinc, caused by the diet high in refined carbohydrates, may be the chief cause of acne.** As we know, all the zinc has been removed from sugar cane when it is refined into white sugar. And all the zinc is removed from white flour and processed cereals when they are refined into the satiny, white, fluffy stuff called white flour and the crispy, crunchy, sugary stuff in the supermarket cereal boxes.

**Zinc is essential for processing carbohydrate foods.** It is vitally involved in regulating blood sugar levels which are usually deranged by diets high in sugar and refined starch. So perhaps it is indeed the lack of zinc, along with lack of the B vitamins (also necessary for the body to process sugar and starch) which brings on acne.

In any case, it's an easy treatment to try for oneself. The Swedish doctors gave 135 milligrams of zinc sulfate. That is, the tablets contained more than 135 milligrams, but they provided that much zinc every day. No mention is made of any change in diet, but it goes without saying that improvement would certainly be noticed much sooner by someone taking the zinc, and also revising meals and snacks so that sugary and starchy goodies are eliminated.

Why not try it, if you are bothered by acne? The diet to follow is very simple: eat only meat, fish, poultry, dairy products, all vegetables and fruits in any quantity, plus real wholegrain breads and cereals, seeds and nuts. And nothing more. No soft drinks. No candy, no desserts except fresh fruit. Zinc is available at your health food store in many preparations. Space the tablets out through the day taking them with each meal, rather than taking them all at once. A chelated zinc preparation will probably make absorption of the mineral more certain. And, of course, continue to take your usual vitamin and mineral supplements each day as, we hope, you always do.

According to a test reported in *American Family Physician* for March, 1971, those foods which for years have been thought to be the leading cause of acne among teenagers have been shown to have no relation to this annoying skin disorder. Chocolate, nuts, cola drinks, milk, cheese, butter, fried foods, iodized salt have all been incriminated in the past as the villains which cause acne pimples. Every new acne patient visiting the University of Missouri Health Service was asked about his own sensitivity to these foods. Those who were convinced that their bad skin was caused by eating one or another of these were then tested for a week by eating considerable amounts of that food under close supervision.

Each individual was given as much as six small chocolate bars daily, a quart of milk, four ounces of peanuts, and so forth, depending on the food to which the patient thought he was allergic. In no case was there any flare-up of the acne.

Everyone was amazed. **It seems obvious that acne is a disease of nutritional deficiency brought on by the very unwise eating habits of our young people.** Surely a diet high in protein, vitamins and minerals, with generous amounts of food supplements and complete elimination of all refined starches and sugars would produce good results. Why not try it?

# CHAPTER 9

# Sickle-Cell Anemia

SICKLE-CELL ANEMIA is a devastating disease in which a large number of red blood cells are deformed to resemble sickles or crescents. The resulting anemia may cause blood clots, heart attacks, liver problems, painful arthritis with fever, ulcers around the ankles, episodes of severe stomach pain and vomiting, plus many kinds of nerve disorders. Almost every part of the body suffers when something is the matter with blood cells.

This is usually an inherited disorder and it is most generally found in black populations, It also occurs along the Mediterranean and in some parts of India. Those with the disease rarely live beyond the age of 40 because of blood clots and infections in various parts of the body.

**It is believed that a single pair of abnormal genes, one from each parent, is responsible for the disease,** according to *The Book of Health.* Defective hemoglobin results in misshaped red cells and an inability of the blood to carry oxygen, producing anemia.

Victims of sickle-cell anemia are usually poorly developed—that is, they do not grow properly and their bones are not normal. Because of this peculiarity, a specialist in trace minerals may have discovered a treatment for this

74

form of anemia which can possibly arrest it or who knows, prevent it.

Dr. Ananda S. Prasad, professor of medicine at Wayne State University in Detroit, has turned up a great deal of evidence on growth abnormalities caused by zinc deficiency. Giving zinc to some people suffering from dwarfism has encouraged additional growth. Dr. Prasad decided to try zinc on sickle-cell patients to see if he might be able to help their growth problems. He did. And in the process he discovered that their leg ulcers healed, their sex organs became more normal.

Dr. George J. Brewer of the University of Michigan carried these studies further by also giving zinc supplements to sickle cell patients. He discovered that he could actually improve the condition of the blood cells. He gave his patients massive doses of zinc acetate. Every four hours, day and night, they took zinc until they were taking a total of 150 milligrams. He got a 10 per cent improvement in hemoglobin (red blood cells), numbers of red blood cells and survival of red blood cells.

"If this had been the sole beneficial effect," he says in an article in a January, 1977 issue of *Medical World News,* "it would be too small to justify therapy of this complexity." But these were not the only effects of the zinc therapy. Ten patients who took zinc had been suffering from an average of about six pain episodes a year. Zinc therapy cut this down to about 2.3 episodes.

Furthermore, **the zinc cut down the number of irreversibly sickled cells, those red blood cells so badly damaged that they cannot be brought back to a healthy state.** But the number of these cells decreased after zinc therapy. The count actually went down from 28 per cent to 18.6 per cent. It is believed that the actual number of sickled blood cells governs the amount of damage that may be done to blood vessels and spleen. So presumably any decrease in the number of such cells would improve the general condition of the patient.

A group of scientists at the University of Michigan headed by Dr. John W. Eaton is looking into another aspect of minerals in relation to sickle-cell anemia. He has found, he says, that sickled cells contain abnormal amounts of calcium. So something appears to be wrong with the way the bodies of these patients are using calcium. This, too, may be related to zinc. Dr. Eaton found that zinc can partially block the entrance of excessive amounts of calcium into the diseased cells.

**Dr. Brewer thinks that zinc benefits the membrane of the sickled cell by somehow opposing the damage done by excessive calcium, but he does not know as yet just how the whole thing works.** At any rate, it looks very promising for future research, which may be able to devise a way to use zinc so that further degeneration of the patient's blood may be arrested. Or, it seems to us, such a trend in research may produce some astonishing insights into what causes this disease. Might the original injury which caused the hereditary condition not be related to a lifetime deficiency in zinc?

We do not know. Undoubtedly any such discoveries are far in the future. Part of the reason why any therapy with zinc is difficult is the way the supplement must be given. Investigators have found that the amount of zinc in an individual's blood rises to a peak about two hours after he takes a zinc supplement. Then it goes down to its former level in about five hours. So the zinc must be taken in very small doses every four hours around the clock, which is a difficult and inconvenient schedule to maintain.

Few patients with a serious case of sickle-cell anemia live beyond the age of ·30 to 40. Infections (tuberculosis, for example), blood clots in the lungs or in some other vital area are usually the cause of death. Any harmless therapy, such as zinc therapy, which can do anything to alleviate the sickling of blood cells which produces the disease is certainly a ray of hope, since no other therapy is known.

Researchers trying to solve one physiological mystery

must, it seems, confine themselves to a study of only one element—zinc in this case. What great advances might be made in understanding this disorder if scientists would use, in addition to the zinc, the best possible diet, as highly nutritious as could be devised, along with supplements of all the vitamins and all the minerals known to be essential!

# CHAPTER 10

# Mental Illness

WE READ RATHER regularly in the medical literature of reports on the treatment of **schizophrenia**—our most serious mental disorder—with massive doses of B vitamins and vitamin C, plus a highly nourishing diet. Now a New Jersey psychiatrist has reported that supplements of zinc and manganese also appear to have a beneficial effect.

As reported in *Medical Tribune*, Dr. Carl C. Pfeiffer of the New Jersey Neuropsychiatric Institute, Princeton, New Jersey, told an international Symposium in Clinical Applications of Zinc Metabolism that "a probable factor in some of the schizophrenias is a combined deficiency of zinc and manganese, with a relative increase in iron or copper or both."

Over half of all hospitalized patients in the United States are mental cases, and about half of those have schizophrenia, according to *The Book of Health*, Third Edition.

**In schizophrenia, the mind turns away from reality into a world of its own creation.** Consequently, the patient's actions are often difficult to understand because they are dictated by the fantasies which rule his mind, the encyclopedia says.

"The disease was formerly known as *dementia praecox*, which means 'a precocious demented state.' While it is true that the disease frequently does appear in early adult life, this is by no means true in all cases, so Bleuler in 1911 advocated the substitution of a new term. He suggested 'schizophrenia' for the reason that 'schizo' (splitting) 'phrenia' (mind) gave some indication of the 'breaking away' of the patient's mind from its normal evaluation of reality," *The Book of Health* continues.

We are also told that **there are many different forms in which the illness, schizophrenia, manifests itself.** However, denial of reality and inappropriate emotional responses are common to all of them. The distorted content of his mind is revealed by the patient's behavior. He may be given to periods of wild behavior in which he breaks up furniture and throws his entire surroundings into disarray. He may rip off his clothes and go naked, or may decorate himself in all manner of fantastic dress. He laughs or cries without due cause and may use a language, consisting of jumbled fractions of words and phrases, which is incomprehensible to others. He may be confused as to his identity and make fantastic claims that he is someone else of wide repute. The actions and mannerisms of the schizophrenic patient appear bizarre and unintelligible when viewed in the light of reality. They are more easily understood when one realizes that they are products of a dream world, erected because the patient cannot perceive reality in a normal way. When personality disorganization progresses too far, the patient no longer can distinguish facts from fancy.

"Schizophrenia cannot be fully comprehended in terms of one disease," *The Book of Health* says. "Rather, it is a set of complex symptoms with manifestations so varied that it has been called 'a group of diseases.' ... Some cases of schizophrenia, but not all, are found in conjunction with an emotional background which would foster the development of withdrawal tendencies. However, many individuals with just as detrimental a background fail to develop the symptoms of schizophrenia. This would suggest that a person's heritage may render him more susceptible to schizophrenia. There are, in fact, some cases of schizophrenia which are difficult to account for on any other basis than that of organic disorder....

"Extensive research into the pathology in the families of schizophrenic patients shows that parental attitudes and interfamily tensions play a major role in the production of schizophrenia. Kanner is one investigator who has done much research on the early histories of schizophrenics. Looking into the childhood of these patients, he determined there was an extremely 'close connection' between parental attitudes and the meaning attached to life experiences by the preschizophrenic child. Particularly the aggressively oversolicitious parent, who must direct all aspects of the child's life, leaving him no privacy of thought, may drive the child into a shell and so begin the practice of habitual recoil. Parents who make too many frustrating demands and show only impatience when their demands are not fulfilled engender lonely antagonism in the child. The child who feels he cannot depend on anyone erects a barrier of reserve to shield himself. He becomes a quiet, docile child, well-behaved until interruption diverts him from the consolation of his fantasies. For a time, he may try to compensate for his lack of adaptability by reading or studying, displaying to the officious adults his industry and knowledge. This is a dangerous symptom, for it is thought to result from a further withdrawal of the potential schizophrenic. The extra social demands which accompany the onset of puberty may prove too great for the youngster and this personality type and with this background. This is why schizophrenia frequently comes on early in life, when the budding adult begins to realize he is unfitted for normal competition in the external world," *The Book of Health* reports.

**Copper, a trace mineral, is excreted very poorly by many victims of schizophrenia,** according to Dr. Carl C. Pfeiffer. And often high blood levels of copper are associated with the disease. **Copper is a zinc antagonist—that is, the more copper you have, the less zinc you are likely to have.** About one-fifth of all the patients he examined had more copper in their blood than they should have and less zinc. Some patients had too much iron in their blood, some had too little.

When the blood levels of copper increased, the disease grew worse. But when Dr. Pfeiffer gave his patients zinc and manganese supplements, copper was excreted and the proper balance between the two minerals was obtained.

**Dr. Pfeiffer was especially enthusiastic about using the zinc-manganese supplement with women and girls suffering from schizophrenia.** Estrogen, the female sex hormone, is also associated with high levels of copper in the blood. In the mentally ill, these high levels of estrogen may actually approximate those of the ninth month of pregnancy, which are abnormally high. By giving the zinc-manganese supplement (manganese should not be confused with magnesium, another mineral), the amounts of copper can be controlled.

The supplement is also valuable in some cases of **mental depression** because "this may herald the onset of schizophrenia." Though Dr. Pfeiffer does not mention it, readers of this book will recall that lack of zinc is closely associated with high and low blood sugar levels in the diabetic state. And low blood sugar (hypoglycemia) is one of the symptoms often found in schizophrenics. Vitamin C is destroyed by exposure to copper, either in the body or in a kitchen utensil. This may be one reason why massive doses of vitamin C have been found valuable for the mentally ill. Although the increased levels of copper in the blood may destroy some of it, enough is left to do all the important work which vitamin C must do in the body.

Testing urine for the presence of a "mauve factor" has become a standard test for schizophrenia among physicians and psychiatrists who use megavitamin therapy in their treatment of this disabling disorder. Dr. Pfeiffer explains why in the December 14, 1973 issue of *Medical World News*.

He reports that 30 to 40 per cent of all schizophrenics excrete in their urine a certain substance which turns a deep ago as 1963, Dr. Abram Hoffer of Canada and Dr. Humphry Osmond of Alabama discovered this fact and published it in medical journals. Opponents of their theory said that the mauve color represented only a reaction to the tranquilizers being given to the schizophrenic patient. But the substance in urine has now been identified and is known to be something which the disordered, unbalanced body chemistry of the mentally ill person is excreting. (The color mauve is a pale bluish-purple).

Dr. Pfeiffer tells of a patient who arrived at his clinic in 1971 suffering from "an unrelenting inferno of mental and bodily suffering." She had suffered over the years from insomnia, loss of reality, attempted suicide, seizures or convulsions, vomiting and difficulty with menstruation. She had been given nerve tests and psychiatric tests and they were normal. She had been hospitalized and tranquilized, all to no avail. Transferring from one hospital to another, she came at last to Dr. Pfeiffer who gave her chemical tests to determine her body's balance of nutrients—vitamins and minerals, and psychiatric tests to determine whether she suffered from perceptive disorders—that is, whether things looked peculiar to her, sounded peculiar, tasted wrong, smelled wrong.

**He treated her with massive doses of pyridoxine (vitamin B6) and supplements of zinc and manganese.** He gave her group therapy and a tranquilizer. He believes, he said, that the food supplements should be given in two doses a day so that they flood the system of the patient.

**This young patient improved with the vitamin-mineral therapy.** When it was discontinued, she relapsed. Returned to this simple therapy, she improved to such an extent that she has been free from convulsions for two years without other medication. She has made up the schooling she missed and is planning to become a doctor. Dr. Pfeiffer points out that she had trouble with knee joints when she began to menstruate. This gave him the clue that she might need the two minerals, since lack of them produces similar troubles in animals.

The mauve chemical which is excreted in some schizophrenics seems to indicate that copper levels are normal in these people, whereas in other schizophrenics who do not have the "mauve factor" copper is lacking. The patient with "malvaria"—the mauve factor—may also have other symptoms: white spots on fingernails, loss of the ability to dream or to remember dreams after waking, a distinctive, sweetish odor on the breath and addominal pain in the left upper side of the abdomen.

**Other symptoms may be:** constipation, inability to tan in sunlight, itching in sunlight, malformation of knee cartilages, joint pains. They may also have anemia, tremor and muscle spasms. They may be impotent or have menstrual difficulties, low blood sugar and an anemia which does not respond to iron but is improved when they are given vitamin B6.

Schizophrenics, doomed perhaps to a lifetime in mental hospitals, can benefit from this new knowledge of the physical, biological basis of their illness. And, says Dr. Pfeiffer, it's quite possible that, following this line of inquiry, we may discover many useful things about people who do not have the symptoms of schizophrenia.

In his most recent book, *Zinc and Other Micro-Nutrients,* which we discuss in greater detail at the end of this book, Dr. Pfeiffer tells the story of an 18-year-old zinc-deficient male who became emotionally unstable due to the stress of majoring in music at a California college. When admitted to a medical center, he was found to be disoriented as to time and place; he was bothered by constant visual and auditory hallucinations. When he did not respond to treatment, he was sent home to South Carolina, where he was admitted to a hospital for further study and treatment. He was given tranquilizers and other drugs but without success.

Tests indicated a vitamin B6 (pyridoxine) deficiency, and an elevated blood pressure suggested an abnormal amount of copper in his system. He was delusional, hallucinating and

83

self-destructive, and, when questioned, he slowly repeated the words of the examining doctor. Shock treatments brought a temporary relief, but, after 10 days, he returned to his former emotional state and tried to jump through a window.

Since everything else had failed, the doctors decided to test the trace metal levels in his blood serum. The zinc was 65 microgram percent (normal is 100 to 120); copper was 185 (normal for males is 100 mcg. percent).

The doctors then administered 160 milligrams of zinc sulfate per day, along with one gram of vitamin B6 each day. In two days, the young man became quiet and he was able to leave his locked room and join other patients in the ward. In addition, his muscle rigidity and tremor lessened. He continued to make progress and within one month was considered back to normal. He was soon discharged.

A follow-up by his doctors one year later found the man back in college and doing well. "His zinc level was still low (75 mcg. percent), but his copper level was normal at 90 mcg. percent. The patient cooperates fully and continues to take his daily dose of zinc and B6," Dr. Pfeiffer reports.

The psychiatrists at the Psychiatric Institute, Columbia, South Carolina, who originally discussed this case history in *Current Psychiatric Digest,* suggest the need to study zinc and copper relationships in psychotic patients. They added that they believe that "there exists a group of patients who have a zinc deficiency which, complicated by emotional stress, may present a schizophrenic picture. Because of the success and safety of the treatment it would seem worthwhile to attempt to identify and treat such patients."

Dr. Pfeiffer's clinic is in Stillman, New Jersey, near Princeton. He is one of the group of dedicated psychiatrists working with the Huxley Institute of Biosocial Research, 1114 First Avenue, New York, New York 10021. If you know someone suffering from the terrible symptoms of schizophrenia, get in touch with the Institute and ask them for literature on this subject. They will be glad to direct you to psychiatrists in your locality, if there are any, who are using nutritional therapy and megadoses of vitamins and minerals, along with more conventional therapy, to treat this disorder.

**Vitamin B12 is definitely a link in the nutritional chain that protects against mental disease**, reports Dr. Roger J. Williams in *Nutrition Against Disease*. In pernicious anemia, caused by deficiencies of this vitamin, the mental symptoms are by no means uniform; they can range from such mild symptoms as having difficulty in concentrating or remembering, to stuporous depression, severe agitation, hallucinations or even manic or paranoid behavior, he says.

"Like the symptoms in pellagra, those caused by vitamin B12 deficiency may be very similar to those observed in schizophrenia," Dr. Williams states. "Yet the relationship between pernicious anemia and B12 is not simple; other factors may be involved as well. Sometimes administering B12 will clear up the mental symptoms associated with pernicious anemia rather slowly—and occasionally, incompletely. The relationship between vitamin and disease is not as direct as in the case of pellagra."

"A subclinical deficiency (one having no apparent clinical symptoms) of niacinamide (a form of vitamin B3) produces depression; and a niacinamide-deficiency disease, pellagra, produces hallucinations and behavioral changes similar to schizophrenia; in fact, some physicians prefer to classify schizophrenia as subclinical pellagra," says Dr. Richard Passwater in *Supernutrition*.

Writing in *Let's Get Well*, Adelle Davis reports that massive amounts of vitamin C have been used without toxicity in a wide variety of illnesses, including certain mental conditions. "One physician gave a 45-year-old woman with schizophrenia 1,000 milligrams (one gram) of vitamin C every hour, and at the end of 48 hours, by which time she had taken 45 grams, she was mentally well and remained so until she died some time later of cancer," she says.

For additional reading on schizophrenia and biochemical individuality, we refer you to *Metavitamin Therapy* and *Body, Mind and the B Vitamins* by Ruth Adams and Frank Murray, published by Larchmont Books, New York.

An innovative textbook for doctors and other professionals is *Orthomolecular Psychiatry* by Drs. David Hawkins and Linus Pauling, W. H. Freeman Co., 660 Market Street, San Francisco, California.

# CHAPTER 11

# Heart Disease

By USING "judicious supplements with trace minerals, vitamins and hormones," a group of physicians from Baltimore, Maryland and Atlanta, Georgia reduced mortality and recurrence of heart attacks in 25 patients whom they treated for six years. There were no new cases of angina (the terrible pain of a heart in distress). None of the patients who had suffered from angina in the past had any more attacks. None of the patients had to be admitted to hospitals for complications of coronary atherosclerosis—which means obstruction of the important heart artery.

What were other benefits of this simple program involving only trace minerals, vitamins and certain hormones? All patients found they could exercise more vigorously and for longer periods of time. All circulatory symptoms seemed to taper off during the six-year period. There were no adverse effects.

These patients were all earlier victims of severe coronary heart disease. They were doing very badly on the usual heart disease treatment. *Medical Tribune* for August 25, 1971 does not state what the program was. Usually it involves drugs of several kinds, including those that "thin the blood"—the anti-coagulants. And usually it involves a diet strictly limited as to animal fats—the anti-cholesterol diet. We are not told how strict was the diet on which these patients were maintained. However, we do know that results were far from encouraging.

So the doctors decided to try some trace minerals, vitamins and hormones. **They gave their patients zinc, copper and manganese, along with "moderate doses of vitamin E and C" and small doses of estrogen and thyroid hormones.**

Usually half of the patients with heart conditions as serious as these can be expected to die within five years. Only one patient of this group of patients died, and none of the other 24 suffered a new heart attack. So apparently the diet and hormone program was eminently successful.

"The specific dietary supplements used were chosen on the basis of previous observations of their roles in cellular metabolism and in heart disease," the researchers said.

It seems impossible that these physicians could have read much in all the extensive material that appears in medical journals (chiefly foreign ones) about the place of vitamin E in the treatment of heart conditions. It seems, too, that they have read little of all the recent discussions about the effectiveness of vitamin C in circulatory conditions, as well as in all-round safeguarding the health of cells and the physiological cement that binds all body cells together. Otherwise they would have surely given larger doses of these two essential vitamins.

Had they consulted Dr. Evan Shute of the Shute Clinic, London, Ontario, Canada, one of the world's greatest experts in the use of vitamin E, he would have told them that individual dosage is extremely important in using vitamin E for circulatory conditions. And he would have told them that massive doses may be necessary. They say they used only "moderate" doses.

Nor do we know what kind of diet these folks were given. It is only within the past few years that many nutrition experts have decided that overuse of sugar is potentially more damaging than overuse of fat in the diet. We are not told whether these heart patients cut down on sugar and avoided refined carbohydrates.

We also wish that the researchers had set up another experiment to run along with this one, in which they gave a much wider assortment of vitamins and minerals along with a nourishing diet, and skipped the hormones. With such an arrangement, we might know more exactly what part was played by diet and supplements and what were the possibly beneficial effects of the hormones.

It isn't what you would call a real triumph for the cause of good nutrition and food supplements in the fight against heart and circulatory disorders—but it's a step in the right direction. It is recognition, on the part of five influential medical researchers, that diet supplementation is not just another idea of food faddists and cranks.

Some fascinating relationships among trace minerals made the headlines at a recent meeting of the Federation of American Societies for Experimental Biology. Trace minerals are those minerals which exist in soil, water and food in such tiny amounts that we speak of them as "traces." They are usually measured in parts per million or even parts per billion.

Enough selenium in soil and water (hence food) seems to promote good circulatory health, according to two scientists from the Cleveland Clinic. They presented maps and statistics which seemed to show that areas chiefly in the Midwest where selenium is abundant in soil and water have considerably fewer deaths from heart disease than areas chiefly along the East and West coasts where there is much less selenium in soil and water.

Texas, Oklahoma, Arizona, Colorado, Louisiana, Utah, Alabama, Nebraska and Kansas have the lowest heart disease death rates. In Colorado Springs the rate is 67 per cent below the national average. But in states such as Connecticut, Illinois, Ohio, Oregon, Massachusetts, Rhode Island, New York, Pennsylvania, Indiana and Delaware, the death rate from heart disease is above the national average, and there is much less selenium in soil and water. Washington, D. C. (low selenium) has a death rate from heart disease which is 22 per cent above the national average. And, incidentally, Washington is a city where liquor consumption is very high.

A second significant finding about selenium is its apparent ability to mitigate the toxicity of another trace mineral, cadmium. Cadmium usually accompanies zinc in nature, so it may be present in more than healthy amounts in areas where zinc smelters or other metal processing factories spew pollution into air and water. Cadmium pollution is increasing in many parts of our country, for other reasons as well. Cigarette smoke contains cadmium. Auto exhaust contains cadmium. Phosphate detergents may contain cadmium, which flows into our waterways along with sewage. Phosphate fertilizers may be a source of cadmium which is sometimes there as a pollutant. Factories which manufacture batteries may release cadmium into nearby waterways as a pollutant. Fish caught in these waters may contain too much cadmium for safe eating. And so on.

Considerable research has shown that exposure to cadmium can cause high blood pressure and related circulatory ills. So perhaps another reason for the good heart and circulatory record of residents of high selenium areas comes from its effect on cadmium in the environment. Zinc is another trace mineral which helps to counteract the ill effects of cadmium. **Zinc, being in the outer coating of cereal grains, is removed wholesale from our cereals and flours when they are refined and processed.** Cadmium, being in the inner, starchy center of the grain, remains there. We get it in white flour, minus the zinc which would help to protect us from its bad effects.

The reason why the fatty substance cholesterol accumulates in unwanted deposits in arteries may have something to do with lack of two minerals in our diets, according to a Univeristy of Cincinnati researcher, Dr. Harold G. Petering. He and his colleagues have been working with laboratory rats to find out more about the connection.

The animals were kept in a laboratory from which all

metals were eliminated, so that no small metallic pollution could influence the results. They were fed diets deficient in zinc and copper. On the deficient diet the rats showed high cholesterol levels in their blood, a condition believed to lead to heart attacks and other circulatory ills. **As the two minerals were added to their diets the cholesterol levels went down.**

No one knows for certain just what these findings may mean to human beings, but Dr. Petering is sure they must be significant, for he says, "Perhaps man is particulary vulnerable to increases in cholesterol, because he has enough zinc and copper in his body to get good growth, but not enough to forestall high and potentially dangerous levels of blood fat.... If man is subjected to certain environmental conditions—such as exposure to a chemical which depresses zinc and copper—he might also get elevated levels of lipids (fats)."

It seems quite possible, does it not, that the "certain environmental conditions" which deprive modern human beings of their essential share of zinc and copper might be the refining of our grains and sugar into commercial cereals, white flour and white sugar. **In both these processes the trace minerals are removed wholesale and they are never restored in the "enrichment" program.** When any mineral appears in a food in appreciable quantity, it means that this mineral is necessary for the human digestive tract to use that food wisely. Minerals and vitamins are built into the very elaborate digestive systems that break down foods into body cells.

If one or more trace minerals are missing, then they must be found elsewhere in the body, or the assimilation process cannot take place. So eating a food entirely devoid of any vitamins or minerals, like white sugar, and, to a lesser extent, white flour makes it very likely that one will soon be deficient in the minerals which originally were in that food.

**Today, half of the calories Americans eat are basically refined cereals and sugar.** So half of our food comes to us deficient in many minerals including the trace minerals. Is it any wonder that heart and circulatory conditions are the leading cause of death? And is it any wonder that these have become the leading cause of death only in the past 40-50 years or so, when our national consumption of these two classes of food has soared to astronomical levels?

In a magazine article on trace minerals, Dr. Jean Mayer, formerly of the Harvard School of Public Health, had this to say about zinc. "It is a key ingredient of enzymes, those substances which help make vital chemical actions take place. For instance, they help move carbon dioxide from the tissues into the lungs so it can be exhaled."

About copper, Dr. Mayer has this to say: "Like iron, it is needed to make red blood cells, but it's needed in much smaller amounts. It's also involved in transporting iron from one part of the body to another and it helps in the formation of bone and brain tissues."

Let's say you want to do something about the new information Dr. Petering has discovered about the possible relationship between zinc and copper deficiency and cholesterol levels. Where should you begin? Can you get a food supplement rich in these two trace minerals? Kelp which is seaweed, is rich in all trace minerals, as sea water is.

Your health food store has zinc and copper supplements. You might consider the chelated products, since the minerals are designed for easy assimilation.

And where, in the food we eat in quantity at mealtime, would we be most likely to find the most copper and zinc? Dr. Mayer lists these foods richest in these two minerals: meat, fish, wholegrains, eggs, fruits and vegetables—those foods which are, or should be, the basis of your diet anyway. **Note that Dr. Mayer points out the necessity for eating wholegrain breads and cereals.**

**"The best place to get zinc,"** he says, **"is from wholegrains.** Like iron, it is lost when grain is refined. But, unlike iron, it is not replaced in 'enrichment' of flour. . . . The need is small, but it must be met."

"Should you eat eggs?" you may ask. "Should you eat meat? Is it possible that we are getting too much cholesterol from these two foods?"

It is true that some scientists and physicians believe one should avoid cholesterol in food, hence avoid meat and eggs. But, as we have seen above, these are two foods which are the best sources of the very trace minerals which are essential to keep us safe from unwanted cholesterol deposits. And, in addition, they are excellent sources of many other nutrients, as well—protein, many vitamins and iron. A few people have inherited a problem with cholesterol, but many nutrition experts now feel that, aside from those people, most of us are more likely to get heart disease from eating too many refined carbohydrates—such as sugar—than from eating foods which contain cholesterol.

**So we have still another bit of evidence of the healthfulness of wholly natural foods like meat, eggs, fish, wholegrains, vegetables and fruits, and yet another claim against refined and processed foods.** It seems the best way to good health is to use freely those foods which we know to be the source of many essential nutrients and to avoid like the plague all those foods which we know to be deficient in almost everything that is needed to nourish human cells. Doesn't this seem to be the best way to avoid not only heart and circulatory troubles, but also lots of other modern maladies as well?

# CHAPTER 12

# Crohn's Disease

An EXTREMELY PAINFUL and serious disorder of the intestinal tract is Crohn's Disease. It is also called regional enteritis or regional colitis or ileitis. It may come on quite suddenly and seem to be an attack of appendicitis. It may become acute at times, then appear to subside, perhaps to appear at a later date.

Like so many unpleasant disorders, we are told that the cause of this disease is not known and that there is precious little that the medical fraternity can do to relieve the pain (often sharp and colicky), to relieve the diarrhea, the ulcers, the adhesions, the cramps, fever and lack of appetite which accompany Crohn's Disease.

**It involves chiefly inflammation of that part of the small intestine which is known as the ileum, the last few feet of the small intestine which join with the colon or large intestine.** All layers of the intestinal wall are involved. The ileum is swollen and red. Ulcers and abcesses may appear. Adhesions may attach one part of the ileum to another part. There may be blockage of the intestine, although no cancerous or tumorous tissues are present. There may be fistulas joining one part of the ileum to another. The disease may progress to peritonitis, which is inflammation of the peritoneum, the membrane lining the abdominal cavity. It may be an acute or chronic condition.

Interestingly enough, **accompanying conditions** may be arthritis (especially ankylosing spondylitis), *erythema nodosum)* (eruptions of tender nodules usually on the legs below the knees, more frequently seen in women);*pyoderma gangrenosum* (ganrenous boils). It sounds like a dreadful health disaster, especially if accompanied by these complications. Because of lack of appetite, the patient may become thin and suffer from malnutrition. There may be bloody stools.

Ileitis may develop after abdominal surgery, after an intestinal infection such as dysentery, from an obstruction, or from irritation of the intestinal lining, reports *Reader's Digest Family Health Guide and Medical Encyclopedia.*

The doctors give mild sedatives and germ-fighting drugs if there are complications. Some form of the cortisone family of drugs may be given to relieve the fever, improve appetite, decrease the number of stools per day and increase intestinal absorption. As a final resort, surgery may be used to cut out the diseased part of the ileum and connect the healthy part to the colon. This is the kind of surgery President Eisenhower had when he suffered from ileitis. He recovered well and lived for some time after the operation.

We recently came upon some new information on this disease which may be helpful in preventing, if not in treating, the condition. Says an article in *Medical Tribune* for September 21, 1977, **"Zinc deficiency is a frequent complication of Crohn's disease and may be the cause of such associated abnormalities as dermatitis, anorexia (lack of appetite), growth retardation and hypogonadism (undeveloped sex organs)."**

Dr. Craig D. McClain, assistant professor of medicine at the University of Minnesota, tested the zinc concentration in the blood of 40 patients, with a follow-up examination of 26 of them six months later. **He found that more than one third of all these patients with Crohn's Disease had levels of zinc in their blood that were too low.**

Four of the 21 men involved (aged 20 to 60) also had

undeveloped sex organs. They were unable to obtain an erection or produce seminal fluid for a specimen. Two of the teenagers involved were considerably smaller than they should have been, another possible symptom of zinc deficiency.

**So far as loss of appetite is concerned, this is a common symptom of Crohn's Disease.** It is believed that part of the reason for this is an abnormality in taste. Food just does not taste as it should to people who are lacking in zinc.

So Dr. McClain tested 10 healthy volunteers with no taste complaints, and seven patients with Crohn's Disease who had plenty of zinc in their blood and seven patients who had deficiency in zinc. There was no difference in taste sensations in the volunteers and the patients with plenty of zinc. **But those patients who lacked zinc in their blood consistently responded unnaturally to taste tests involving salt, sugar, bitter compounds and sour compounds.**

This is only one bit of evidence to add to the mountain of evidence we have assembled in this and other books pointing to lack of zinc in many conditions of ill health.

Many of these disorders of zinc deficiency are part of the welter of "diseases of civilization" which, many experts believe, are caused by what we have done to our food in processing and refining it. We know, for example, that fiber in the diet is essential to prevent many disorders of the colon. Doesn't it seem probable that a disease of that part of the small intestine which leads into the colon could also be caused by lack of fiber? Fiber is the indigestible part of cereals, seeds, fruits and vegetables. So mightn't enough fiber in the diet prevent Crohn's Disease?

Then, too, **zinc is one of the precious food elements that is lost when sugar, cereals and flours are refined.** These foods are among our best sources of this trace mineral. Refining removes all the zinc. It is never replaced. So a generation which grew up on a diet of which one half consists of white flour and white sugar products is bound to be short on zinc.

It all ties together somehow. Lack of zinc occurs for the same reason as lack of fiber. The various woes described in this book are all prevalent disorders in countries where refined and processed carbohydrates form the mainstay of the diet.

Zinc deficiency is found in all of these disorders, including arthritis. And arthritis is often a symptom accompanying Crohn's Disease. Throughout a great deal of research on arthritis, we have found colon and intestinal troubles mentioned frequently as conditions which accompany arthritis. Couldn't the reason for both of them, or at least one reason for both of them, be lack of fiber at meals, resulting in a shortage of all those nutrients which accompany fiber in sugar and cereals—zinc being one of them.

We found one other helpful note on Crohn's Disease. Victims are usually short on vitamin C as well. Whether it is the inflammation involved in this unpleasant disorder, which uses up the vitamin C in the patient's body, or whether it is just that people with Crohn's Disease have not been getting enough of the vitamin to prevent the disease we do not know. But we do know that people with Crohn's Disease have less vitamin C in their blood than well people, and the worse the condition gets, the less vitamin C is in the blood.

Vitamin C is essential for the body to manufacture collagen. This gluey substance is necessary for holding our cells together and repairing any sores or ulcers that exist. The vitamin is used up in the process. If a patient with Crohn's Disease is put on a "bland" diet, he is generally told not to eat fruits or vegetables which are the only reliable food sources of vitamin C, so the situation becomes worse. The information on vitamin C comes from an article in *Gastroenterology* for September, 1974.

Writing in *Food Facts and Fallacies*, Dr. Carlton Fredericks and Herbert Bailey report that a man with a history of allergy or Crohn's Disease might require amounts of the B vitamin pantothenic acid larger than the amounts included in over-the-counter liver concentrates. The authors were discussing the nutritional benefits of liver concentrates, wheat germ, brewers yeast, etc.

# CHAPTER 13

# Epilepsy

THE WORD "epilepsy" means "seizure." Epileptics often suffer from a loss of consciousness, momentary or prolonged, and involuntary, convulsive movements. An epileptic seizure, or fit, is the result of a temporary disturbance of the brain impulses. No one knows exactly why epilepsy occurs, although it can be caused by brain damage or defects, poor nutrition, infectious diseases and other factors, according to *Reader's Digest Family Health Guide and Medical Encyclopedia.*

"Minor seizures, called *petit mal,* last only five to 20 seconds, and the loss of consciousness is momentary," the encyclopedia continues. "Although the victim suffers a twitching about the eyes or mouth, his posture does not change and he appears to have had no more than a moment of absentmindedness.

"In major seizures, or *grand mal,* the victim falls to the floor unconscious for a moment or more, often foaming at the mouth, biting and shaking his limbs violently. Involuntary bowel movements or the passage of urine may occur. The person may hurt himself during such a seizure. Fortunately, people with epilepsy frequently experience a warning, called the aura, before a major attack occurs, and this enables them to lie down to avoid falls."

We also learn that ordinary epilepsy is also called genuine

epilepsy or idiopathic epilepsy, which means the cause is unknown. Epilepsy usually begins early in life and is not directly inherited, although a predisposition to it may run in families.

**Epilepsy may be triggered by hypoglycemia**, as reported in our book, *Is Low Blood Sugar Making You a Nutritional Cripple?* In such cases, a high-protein, low-carbohydrate diet is recommended.

Two nutritional substances were linked recently to possible influence on the course of epilepsy, the brain and nerve disorder which disables many Americans. A Montreal, Canada group of researchers disclosed at a meeting of the American Neurological Association and the Canadian Congress of Neurological Sciences that **one of the amino acids, taurine—a non-essential one—and zinc appear to be related to one's susceptibility to epileptic seizures.**

They do not, as yet, have definite information as to just how these two substances may function, but they are proceeding with further work along these lines. Amino acids are the building blocks of protein—the basic stuff of which we are made. Most of our interest centers on those which we call "essential", meaning that we must get them in food, since our bodies cannot manufacture them. But taurine is a non-essential amino acid. By that we mean that the normal body can make it so there is no need to get it in food.

So how could anyone be deficient in taurine, if indeed we can make it ourselves without the necessity of getting it in food? One presumes that something in the epileptic's physiological make-up may prevent him from making his own taurine. In that case, giving the amino acid might repair the damage. Of course, it would have to be given for the rest of his life.

Working on this hypothesis, Dr. John Donaldson and Dr. Andre Barbeau of Montreal's Clinical Research Institute gave taurine to 12 epileptic patients who were having at least three seizures a day, although they were all taking maximum doses of conventional medicine. The seizures decreased in frequency within 24 hours and later were eliminated entirely, although the patients are still getting their anticonvulsant drugs. The doctors are not sure what level of the amino acid should be given, so they cannot suggest dosages for others. It is too early to tell.

The other nutrient which they feel may be involved with epilepsy is zinc—the trace mineral which is necessary for preventing many other conditions of ill health. The reasons why zinc may be important for this purpose are too complex for explanation to the layman. Basically they have to do with the possibility that **zinc may be involved in binding a certain substance in a certain part of the brain so that it is there to perform its function. It is well known that the amount of zinc in that part of the brain is considerable.**

Looking further, the two doctors discovered that 32 of their epileptic patients had 15 per cent less zinc in their blood than non-epileptics. It seems to them that this trace mineral may work along with the amino acid to provide what is lacking in the make-up of the epileptic.

Both doctors emphasize that their findings are very preliminary and they can give no definite answers as yet. But it looks hopeful. **Meanwhile, doesn't it seem possible, even quite likely, that epileptics just don't get enough zinc at their daily meals?** Zinc is not the easiest mineral to come by in modern meals, since seeds and wholegrain cereals are among its most abundant sources.

Since the Montreal doctors' research was reported in *Medical World News* for September 7, 1973, it is possible they have made more significant discoveries about epilepsy.

*The Lancet* for February 24, 1968 discusses **schizophrenia and epilepsy as opposite sides of the same coin— that is, biologically antagonistic diseases**. The author gives extensive references to earlier papers that have shown that epilepsy does not appear among people who are schizophrenic, and that, on the contrary, a considerable number of people who suffer from schizophrenia also suffer from epilepsy.

In addition, it seems that **an anti-convulsant drug given to epileptics can bring on a condition of schizophrenia and, more recently, the discovery that tranquilizers given to schizophrenics can precipitate epileptic attacks**.

It never seems to occur to medical researchers that the condition that causes schizophrenia might well be the same condition that causes epilepsy. Does it not seem possible that the drugs given to correct the schizophrenia cause low blood sugar, which produced the epilepsy and that the drugs given to prevent epilepsy somehow so twisted and perverted the mechanism for delivering sugar to brain and nerve cells that schizophrenia resulted! Why is it impossible for doctors to see that both conditions may be simply manifestations of blood sugar disorder as are diabetes and hypoglycemia?

Why must most doctors set diabetes in one category and deny that there is any category of disease resulting from the opposite of diabetes? Why must they decide that epilepsy is one set of conditions having nothing to do with diet and schizophrenia is another set of conditions having nothing to do with diet? Why not, in all four instances, try first to treat the conditions with the same sensible, highly nutritious diet and see what the results will be?

The mechanism that is disordered in all four conditions involves the immeasurably complex process by which sugar is fed to brain and nerve cells. They must have a constant supply of sugar (glucose) or they will malfunction. Is it not possible that the way in which sugar is kept from the brain and nerve cells in different individuals may determine whether they get epilepsy or schizophrenia, diabetes or low blood sugar, the four conditions we are talking about? Individuals differ in their response to stress and lack of sugar in brain and nerve cells is stress.

**"The sugar-laden American diet has led to a national epidemic of hypoglycemia,** an ailment charcterized by irrational behavior, emotional instability, distorted judgment and nasty personality defects. Almost 10 per cent of the population is hypoglycemic," state Dr. E. Cheraskin, Dr. W. M. Ringsdorf, Jr. and Arline Brecher in *Psychodietetics*, published in 1974.

An editorial in *Medical Tribune* for August 24, 1977 blows the whistle on sugar, or sucrose as the doctors call it. No question about it, says the editorial, **sugar in the amounts in which we Westerners eat it, is turning out to be one of the most dangerous things in our environment.**

"The findings come from experiments on human subjects as well as from studies of several species of animals utilizing the amount of sugar comparable to those taken by many people in the United States and other Western countries," says the *Tribune*.

In animal experiments, biologists have found that sugar can produce enlargement of the liver and kidney, can cause considerable alteration of the enzyme activity in the liver, the kidney and the fatty tissues of the body, can interfere with the body's utilization of the protein in meals, can cause changes in the tissue of the retina of the eye and the male reproductive organs.

"These experiments," says the *Tribune*, "suggest involvement of dietary sucrose in human disease." The understatement of the year, perhaps.

One group of highly qualified scientists believe that **sugar, not saturated fat, is responsible for our epidemic of heart disease.** The sugar we eat in such vast, unnatural amounts leads to an increase in the fats in the blood, also leads to increased uric acid in blood (the cause of gout), also leads to disorders of blood sugar regulation, increased adhesiveness of certain blood cells (which might produce strokes or heart attacks), leads to higher levels of insulin in the blood and higher levels of cortisol (a hormone produced by the adrenal glands).

And, furthermore, these same well-qualified scientists and medical men believe that **sugar, in the immense and un-natural amounts in which we eat it, may be one of the** causes of diabetes, especially the diabetes that appears in later years, not only because these large amounts of sugar have a profound effect on blood sugar but also because they may prevent insulin from entering cells as it is meant to.

Recent work has concentrated on the kidney, says the *Tribune*, because it is believed that eating sugar may contribute to the kidney complications of high blood pressure and because eating sugar results in kidney enlargement. Scientists want to study the tissues of these enlarged kidneys and find out just how and why sugar produces these complications.

The *Tribune* feels it is good that these aspects of sugar—over-consumption are coming to light, for they may help doctors to understand more about the heart problems that plague diabetics. These findings may be "of major significance."

The health food industry cannot say "we told you so" because it would be too cruel a thing to say to the millions of people afflicted with the many disorders which are listed in just this short editorial: disorders of liver, kidney, blood sugar, adrenal glands and pancreas, protein utilization, eyes, reproductive organs and much more.

But we can say that the health food movement should take credit for being the first group in the country to warn of excessive sugar consumption. We have been doing this for almost a century. We have been treated with scorn and contempt, laughed at and persecuted. We have remained steadfast in our position that sugar, in the amounts in which we Westerners consume it, is one of the greatest threats to our national health. Now that medical science and laboratory science are reluctantly and slowly coming around to our point of view, it seems possible that we may be able to save the people of the future from making the grievous mistakes the people of our century have made in regard to over-consumption of sugar. We can do little for those millions already irreparably damaged by sugar.

We should do everything in our power to alert government officials to the terrible threat sugar poses to our national health. To that end, we suggest you convey to your representative and/or senators in Washington, some of the material on sugar presented above along with some comment on the following quote from *The New York Times*, which indicates how our Congress views the potential health hazards of sugar.

"House and Senate conferees agreed today on a new price support program for sugar that would benefit producers but would cost consumers and big industrial users $440 million and $660 million a year. . . . The Carter administration, with the reluctant support of the Agriculture Department, had proposed a one-year program of direct subsidies to sugar producers that would have raised their income to 13.5 cents a pound. This has been highly favored by the large sugar users, such as the Coca-Cola Company, because their costs for sugar would have stayed at the lower free-market prices while the taxpayers picked up the $480 million sugar subsidy bill," stated *The New York Times*, August 8, 1977.

In the meantime, getting back to the subject of this book—zinc—we know that this mineral is closely tied in with the body's use of insulin. This directly or indirectly may involve diabetes and low blood sugar, which are also associated with epilepsy, schizophrenia and a host of other disorders.

The addition of zinc to insulin solutions given to diabetics delays the action of insulin, so that the diabetic has a longer period of lowered blood sugar, hence does not need more insulin quite so soon.

Before we leave epilepsy and schizophrenia, Dr. Abram Hoffer, writing in the October, 1975 issue of the *Huxley Institute-Canadian Schizophrenia Foundation Newsletter*, says that there are people who need far more pyridoxine (vitamin B6) than the rest of us. He believes that as many as 5 per cent of any "normal" population and up to 75 per cent of a severely ill psychiatric population have this problem. They need more pyridoxine than they can get in the average diet.

"As well as being mentally ill," he says, "these patients may have constipation and abdominal pains, unexplained fever and chills, morning nausea, especially if pregnant, hypoglycemia, impotence or lack of menstruation, and neurological symptoms such as amnesia, tremor, spasms and seizures. The treatment consists of optimum amounts of B6 and zinc." About one-third of all schizophrenics suffer from this need for very large amounts of the vitamin.

# CHAPTER 14

# Boils

A SWEDISH PHYSICIAN has found, he says, a cure for boils. He reported on his work in the *Lancet* for December 23-31, 1977. He treated fifteen patients from 24 to 50 years of age. **Testing their blood, he found that all of them were deficient in zinc.**

These people had suffered from boils (furunculosis doctors call them) for 3 to 10 years. The boils appeared on the groin, thigh, buttocks, abdomen, breasts, face and neck, as often as once or oftener every month. The boils often appeared where clothing had rubbed against the skin.

In seven of the patients Dr. Isser Brody, a Swedish dermatologist, used the orthodox method of treatment. He opened the boils and prescribed antibiotics. The patients returned to the office regularly for three months and reported that they had new boils which the doctor then had to open and treat. The levels of zinc in their bodies remained low during this time. The doctor did nothing about this except to measure the zinc every time they came in.

He gave the other eight patients zinc sulfate **as the only treatment they were to receive.** They took three tablets of this preparation every day (one with each meal). Each tablet contained 45 milligrams of zinc. The doctor asked these patients to return and report to him.

Seven of them returned at intervals for three months. One

patient came back frequently for seven months. In all of the eight patients the zinc levels in their blood rose to normal within a month. **All boils disappeared and no new ones appeared with no treatment but zinc supplements.**

Here is a case where we wish the dermatologist had bothered to get a complete history of eating habits of these patients. It seems incredible that he never wondered, apparently, why they should have low levels of zinc in their blood. Doctors in general seem to feel that diseases fall from heaven upon mortals for no reason and that what you eat has nothing to do with what disease you have.

Had Dr. Brody inquired about the meal patterns of his patients we could probably discover why they suffered from boils all their lives. **Here are some of the possible medical reasons for zinc deficiency** as outlined by Ananda S. Prasad in his classic book *Trace Elements in Human Health and Disease, Vol. 1, Zinc and Copper:* alcohol, excessive sweating, blood loss, malabsorption (that is, some intestinal condition which prevents absorption of minerals) prolonged intravenous fluids, fasting, infection, nephrosis (kidney disease) chelating agents in the form of drugs, heart attack, surgery, cirrhosis of the liver, certain hormones given to patients, pregnancy, lactation and certain inherited characteristics.

Dr. Prasad also mentions phytate, the substance in whole grains which tends to deplete the body of many minerals, including zinc, when *unleavened* bread is eaten. Bread raised with yeast does not cause this depletion.

**Here are some other possible causes of zinc deficiency:** disorders of the pancreas, partial or total removal of the stomach, diverticuli of the intestine, dialysis, certain kinds of anemia, cancer, psoriasis, burns, intestinal worms, diabetes, absence of the thymus gland, mongolism and a skin condition called acrodermatitis enteropathica.

**Other reasons, of course, are dietary habits which exclude those foods containing the most zinc.** Dr. Henry Schroeder tells us, in his book *Trace Elements and Man* that the milling of wheat into refined white flour removes the following trace minerals. None of these except iron, is replaced in the white flour: 40% of the chromium, 86% of the manganese, 76% of the iron, 89% of the cobalt, 68% of the copper, *78% of the zinc* and 48% of the molybdenum.

So the person who has eaten all his or her life only bread made from white flour has lost most of the zinc that accompanies that flour in its natural state. **We need the zinc in order for our bodies to process the carbohydrate of the flour.** It's not there. And most supermarket breads which claim to be wholegrain are made mostly from white flour with just a bit of wholewheat added.

In refining sugar from sugar cane, 93% of the chromium is removed, 89% of the manganese, 98% of the cobalt, 83% of the copper, *98% of the zinc*, and 98% of the magnesium. We know that zinc is essential for processing sugar in the body. Zinc is concentrated in the pancreas which is the gland most concerned with processing sugar. It's not present in white sugar.

Statistics tell us that about one half of the diet of most Americans consists of these two foods—white sugar and white flour and foods made from them. Says Dr. Schroeder, "Most of the energy in the average American diet, which comes from white flour, white sugar and fat, is not supplied with the trace substances needed to utilize that energy efficiently and properly." Zinc is one of these. The same is, of course, true in European countries where the same kind of diet prevails.

**So you might over-simplify and say that boils like many other skin diseases come from eating too much sugar.** Because sugar and white flour crowd out of the diet those nourishing foods which are rich in trace elements like zinc and chromium and because the giant food industry has removed these trace elements wholesale from our basic staple foods, diseases occur which are caused, apparently, by simple lack of these trace minerals.

**Can you cure your boils with zinc?** There's no guarantee, of course, but there's no harm in trying. Zinc supplements are available. According to Dr. Schroeder up to 150 milligrams of zinc daily are well tolerated by human beings. No one is quite certain what the toxic dose might be. If you decide to try zinc supplements to treat your boils, don't for goodness sake do it without at the same time eliminating sugar and white flour from your meals, for you are simply pouring zinc into your body through pills, and depleting your body of zinc by continuing to eat these two pernicious, demineralized foods. What's the point?

Here are some other brief notes on zinc therapy for different ailments. *The Lancet* for October 9, 1977 reported on eight men with kidney disease so serious they were on dialysis machines. All were impotent. They had extremely low levels of zinc in their blood. Their doctors at the Georgetown University School of Medicine in Washington, D.C. gave them zinc supplements through the dialysis machine and "strikingly improved potency in all patients" and improved as well the condition of sex hormones in two patients.

In a press release on December 14, 1977, researchers at the Massachusetts Institute of Technology reported that throat cancer, which claims 6,500 lives a year in this country, may be triggered by low levels of zinc. These physicians have observed low levels of zinc in the esophagus of victims of this terrible disease. They believe that lack of zinc may sensitize the esophagus, making it more susceptible to cancer. After making this discovery, the scientists gave one group of rats a diet deficient in zinc, another group a well-balanced diet. Then they injected a chemical known to cause esophageal cancer. The rats deficient in zinc developed more cancers in a shorter period of time than those animals which had gotten the good diet, rich in zinc.

Say these scientists, "We don't know why the chemical should seek out the esophagus. Nor do we know exactly how esophageal cancer develops. But we know that (lack of) zinc is somehow involved, along with smoking, consumption of alcohol, food contaminants and perhaps other environmental factors in causing cancer."

A short note to the editor of *The Lancet*, November 26, 1977 details much information on **the influence of zinc levels on male sex life.** Three European doctors report results from giving zinc supplements to ten infertile men suffering from lack of viable sperm, as well as deficiency in certain male hormones. The doctors gave zinc sulfate in 220 milligram doses three times daily for 4 to 8 weeks. The zinc levels in blood and in male hormones were raised, as well as sperm count. Another 15 patients treated later seem to show the same results. The wife of one patient became pregnant after her husband had taken zinc for six weeks.

Say these researchers, "Zinc plays an important role in prostatic, epididymal and testicular function, but the relationship between (deficiency in sperm) and zinc has not yet been studied. Our findings seem to open a new prospect for the treatment of some cases of deficient sperm of unknown origin."

*The Lancet* for November 19, 1977 reported on effects of zinc deficiency in malnourished children. It causes their thymus glands to atrophy and produces increased susceptibility to infection.

*New Scientist* for December 1, 1977 published a major article by a biochemist, Dr. Robin Wilson, entitled *Zinc: A Radical Approach to Disease.* Says Dr. Wilson, "Because the metal (mineral) content of food is often lowered during processing (refining flour, for example, can reduce both the iron and zinc content of whole grain) and because people are eating more processed foods, there are grounds for concern that the average diet may be becoming zinc deficient. A consensus of dietary studies indicates that an adult needs to absorb 5-7 milligrams of zinc a day compared with 1-2 milligrams of iron. On average, approximately one-third of the zinc in food, but only 10 per cent of the iron, is absorbed. Thus the intake of both metals should be in the region of 15 milligrams a day."

**He goes on to speculate how much zinc may be lost**

**from drinking alcohol, from infections and so on.** Iron is added to white bread in England, as it is in our country. **But zinc is neglected.** It is estimated that only about one per cent of added iron in bread is absorbed, compared to 20 per cent absorption of iron from meat. But, says Dr. Wilson, so many foods are now being fortified with iron that we may be getting more than twice the accepted daily iron requirement. A small bowl of one Kellogg's breakfast cereal, for example, contains as much as 12 milligrams of iron "powder"—that is, the iron is added, after most of the iron naturally in the cereal has been removed by processing.

Until recently, says Wilson, only two body enzymes were known to contain zinc. **Over 80 have now been identified.** This indicates that 80 essential body processes are governed by zinc. If you are deficient in this mineral, all these processes will suffer and not function correctly. Because of the curious relationship between iron and zinc, it seems possible to Dr. Wilson that an excess of iron added to processed foods may be resulting in a greater than normal deficiency in zinc . . . It is possible, he says, **"that many more of us may be consuming too much iron and not enough zinc."**

Best sources of zinc are all fish and shellfish, liver, meat, eggs, (just the yolk), whole grains, all seed foods such as nuts, peanuts, lentils, soybeans, peas, corn, brewers yeast, rice, oatmeal, milk are also good sources. It's easy to see how someone frightened of cholesterol might eliminate from his diet all zinc-rich shellfish, liver, eggs, and meat (they are all rich in cholesterol) and depend on white bread, supermarket cereals and sugary desserts—the very foods most completely deficient in zinc.

If you wish to take zinc supplements, there is no reason to take zinc sulfate, as the volunteers did in the experiments we have described. Many zinc supplements are available at your health food store. The zinc in the ones marked "chelated" is much more likely to be fully absorbed.

# CHAPTER 15

# Do You Know Your Trace Minerals?

"TRACE ELEMENTS ARE the nutrients man evolved on, abundant in our diets and those of our ancestors for all but the last twenty of the many millions of years it took life to evolve. In these past twenty years, food processing and high-yield farming techniques have caused depletion in the amounts of these nutrients found in our foods," says Carl C. Pfeiffer in a new paperback *Zinc and Other Micro-Nutrients*.

**Dr. Pfeiffer is talking about the beneficial trace minerals such as zinc, manganese, selenium and chromium.** Just as important to consider, for quite another reason, are the highly toxic trace minerals like cadmium and lead. Modern engineers have discovered they can use many such poisonous trace elements in the manufacture of many new produc . So they are mining them and using them extensively.

Cadmium, for example, is poisonous enough to be an ingredient in a weedkiller. Yet up until only a few years ago, human beings had almost no exposure to this trace mineral, so, of course, they could develop no protective bodily mechanism to guard against it. Yet today, cadmium is in every puff of cigarette smoke. It is a pollutant in galvanized piping used in old buildings. People living downwind from zinc smelters run the risk of cadmium poisoning since this pollutant is ever present in the air. A factory discharging cadmium waste into a lake in Japan brought about the poisoning of almost all the nearby residents and the deaths of more than 100. These folks used the lake water to irrigate their rice. Over the years poisonous levels of cadmium built up in the rice.

**The Pfeiffer book brims over with helpful suggestions on how best to use the beneficial trace minerals for good health and how to counteract the evil effects of the toxic ones.**

For example, chromium has recently been discovered to be essential for regulating blood sugar levels. Yet all of it is removed from grains when they are processed and refined to make white flour and boxed supermarket cereals. Could this not be one of the main reasons why figures on the incidence of diabetes are soaring? Chromium and zinc have been found to be essential for the health of the eyes. Both are also removed in grain and sugar processing. So we have an epidemic of eye troubles: cataract, retinal detachment, glaucoma and degeneration of the cornea in older folks.

Mysterious incidents have taken place involving the trace mineral cobalt. We need cobalt, for it is an integral part of vitamin B12 and has no other use in the body, so far as we know. But several years ago an epidemic of deaths from sudden heart attacks occurred among heavy beer drinkers. It took careful detective work to piece together the story. It seems that synthetic detergents leave a film on drinking glasses which keeps down the froth on beer. So to encourage the froth to stand high, breweries were using cobalt in relatively large amounts in their beer. The amount of cobalt heavy beer drinkers were getting was enough to poison them and cause fatal heart attacks.

The Pfeiffer book contains many such stories on both the harmful and the beneficial trace minerals. In straightens out the complicated relationships among them, too. It tells you why and how we must balance calcium and phosphorus as well as sodium and potassium. It stresses the need for getting enough of these precious trace elements which have been almost completely removed from those foods which provide the staple, basic diet of most people—the refined carbohydrates.

For example, the trace mineral chromium is part of a "factor" which regulates blood sugar levels, hence is absolutely essential for preventing diabetes. Yet chromium is removed wholesale from those foods which, it is now believed, bring on diabetes—the foods that are loaded with sugar. So the diabetic has to fight not only the ravages of the disease, but also the deficiency in chromium which further devastates his health.

**Zinc is essential for the proper activity of the pancreas which is disordered in diabetes.** Yet no doctors give zinc to diabetics. Why not? Zinc, too, has been removed from all refined carbohydrates. Manganese, another trace mineral, has also been found to be related to blood sugar regulation. It is removed wholesale from white flour and white sugar.

Calcium, a mineral which has been studied thoroughly for many past years, is now known to be related to many other minerals and trace minerals in very complex ways. Says Dr. Pfeiffer, "Calcium nutrition is a complex matter which must be regarded as a whole rather than discussed in a fragmentary manner. Dr. William Strain of Cleveland has stated that trace element nutriture is like a giant spider web; if one branch of the web is pulled, the whole web of trace elements becomes distorted. Calcium balance, so closely related to magnesium, zinc, iron, selenium and sulfur balance, exemplifies this analogy very well."

That is the stuff of which this book is made. It is engrossing and easily understood. Most of all it is helpful. Anyone who reads it is bound to discover scores of helpful suggestions on regulating his own diet and diet supplements.

The name of the book is *Zinc and Other Micro-Nutrients* by Carl C. Pfeiffer, a paperback published by Keats Publishing, Inc., 36 Grove Street, New Canaan, Conn. 06840 for $2.25.

# Index

119

# The Larchmont
# PREVENTIVE HEALTH LIBRARY

The **Preventive Health Library** has been designed to give you all the facts concerning the uses and effects of a particular supplement in a handy, easy-to-read format that is highlighted to help you digest key concepts, All books are by Adams and Murray, the famous health team who keep you informed on the most valuable health topics.

**Improving Your Health with Vitamin A.** 128 pages, $1.25.

**Improving Your Health with Vitamin C.** 160 pages, $1.50.

**Improving Your Health with Calcium and Phosphorus.** 128 pages, $1.25.

**Improving Your Health with Vitamin E.** 176 pages, $1.50.

**Improving Your Health with Niacin (Vitamin B₃).** 128 pages, $1.25.

**Improving Your Health with Zinc.** 128 pages, $1.50

*The best books on health and
nutrition are from*

# LARCHMONT BOOKS

—"**The New High-Fiber Diet,**" by Adams and Murray;
foreword by Sanford O. Siegal, D.O., M.D.; 320 pages,
$2.25

—"**Program Your Heart for Health,**" by Frank Murray;
foreword by Michael Walczak, M.D., introduction by E.
Cheraskin, M.D., D.M.D.; 368 pages, $2.95.

—"**Food for Beauty,**" by Helena Rubinstein; revised and
updated by Frank Murray, 256 pages, $1.95.

—"**Eating in Eden,**" by Ruth Adams, 224 pages, $1.75.

—"**Is Low Blood Sugar Making You a Nutritional
Cripple?**" by Ruth Adams and Frank Murray, 176 pages;
introduction by Robert C. Atkins, M.D.; **$2.25**

—"**Beverages,**" by Adams and Murray, 288 pages, $1.75.

—"**Fighting Depression,**" by Harvey M. Ross, M.D.; 224
pages, $1.95.

—"**Health Foods,**" by Ruth Adams and Frank Murray,
foreword by S. Marshall Fram, M.D.; 352 pages, $2.25.

—"**Minerals: Kill or Cure?**" by Ruth Adams and Frank
Murray; foreword by Harvey M. Ross, M.D.; 368 pages,
$1.95.

—"**The Compleat Herbal,**" by Ben Charles Harris, 252
pages, $1.75.

—"**Lose Weight, Feel Great**," by John Yudkin, M.D.; 224 pages, $1.75.

—"**The Good Seeds, the Rich Grains, the Hardy Nuts for a Healthier, Happier Life**," by Adams and Murray; foreword by Neil Stamford Painter, M.D.; 304 pages, $1.75.

—"**Megavitamin Therapy**,' by Adams and Murray, foreword by David Hawkins, M.D.; introduction by Abram Hoffer, M.D.; 288 pages, $1.95.

—"**Body, Mind and the B Vitamins**," by Adams and Murray, foreword by Abram Hoffer, M.D.; 320 pages, $2.75

—"**The Complete Home Guide to All the Vitamins**," by Ruth Adams, foreword by E. Cheraskin, M.D.; 432 pages, $2.95.

—"**Almonds to Zoybeans**," by "Mothey" Parsons, 192 pages, $1.50.

—"**How To Control Your Allergies**", by Robert Forman, Ph.D., foreword by Marshall Mandell, M.D.; 256 pages, $2.50

—"**The Vitamin B-6 Book**", by Adams and Murray; 176 pages, $1.75

---

"**LARCHMONT BOOKS are available from health food stores in your area. If your store does not stock the title for which you are looking, please ask the store owner whether he or she might order it for you.**"

**Read What the Experts Say
About Larchmont Books!**

**The Complete Home Guide
to All the Vitamins**

"This is a handy book to have at home, for it discusses in clear, simple language just what vitamins are, why we need them, and how they function in the body."—*Sweet 'n Low*

"Want to know what vitamins you need and why? Then this is your cup of tea. A paperback that tells you everything you ever wanted to know about vitamins and maybe were afraid to ask...Read it and reap."—*Herald American and Call Enterprise, Allentown, Pa.*

**Minerals: Kill or Cure?**

"Written both for professional and non-professional readers, this book offers excellent background for additional discoveries that are inevitable in the next few years..."—*The Total You*

**Eating in Eden**

"This book contains very valuable information regarding the beneficial effects of eating unrefined foods..."—*Benjamin P. Sandler, M.D., Asheville, N.C.*

"We must be reminded again and again what junk (food) does and how much better we would be if we avoided it. This book serves to do this."—*A. Hoffer, M.D., Ph.D.*

# Read What the Experts Say About Larchmont Books!

## Megavitamin Therapy

"This book provides a much-needed perspective about the relationship of an important group of medical and psychiatric conditions, all of which seem to have a common causation (the grossly improper American Diet) and the nutritional techniques which have proven to be of great benefit in their management." —*Robert Atkins, M.D., author of "Dr. Atkins' Diet Revolution," New York.*

"This responsible book gathers together an enormous amount of clinical and scientific data and presents it in a clear and documented way which is understandable to the average reader . . . The authors have provided critical information plus references for the acquisition of even more essential knowledge."—*David R. Hawkins, M.D., The North Nassau Mental Health Center, Manhasset, New York.*

## Health Foods

"This book (and "Is Low Blood Sugar Making You a Nutritional Cripple") are companion books worth adding to your library. The fact that one of the books is labeled "health foods" is an indication how far our national diet has drifted away from those ordinary foods to which man has adapted over the past million years. . . ."—*A. Hoffer, M.D., Ph.D., The Huxley Newsletter.*

"A sensible, most enlightening review of foods and their special qualities for maintenance of health . . . ."—*The Homeostasis Quarterly.*

## Read What the Experts Say About Larchmont Books!

### Body, Mind and the B Vitamins

"I feel that "Body, Mind and the B Vitamins" is an excellent, informative book. I recommend everyone buy two copies; one for home and one to give to their physician."—*Harvey M. Ross, M.D., Los Angeles, Calif.*

### Program Your Heart for Health

"What is unique about this book is that the tremendous body of fascinating information has been neatly distilled so that the problems and the solutions are quite clear.... (This book) will be around for a long time ... so long as health continues to be the fastest growing failing business in the United States and so long as it is not recognized that the medical problem is not medical but social."—*E. Cheraskin, M.D., D.M.D., Birmingham, Ala.*

"If more people were to read books such as this one and were to institute preventive medical programs early in life, the mortality in heart disease would drop precipitously as well as in our other serious medical problems."—*Irwin Stone, Ph.D., San Jose, Calif.*

"**Program Your Heart for Health**" contains a wealth of data. I plan to make use of it many times."—*J. Rinse, Ph.D., East Dorset, Vt.*